Experiences and Passions

Dick Hyatt

Table of Contents

Dedication

To the memory of Jimmy, James Albert Funk, Jr., for all that we shared together.

Acknowledgments

I want to first give thanks and appreciation to those who urged me to write these stories, and especially to James Weeks who finally got me to start writing. Thank you, James, for not giving up on me.

When I wrote these stories, I rapidly put my thoughts down on paper giving little attention to proper punctuation and even sentence structure. I have had a great deal of help from others who suggested re-arranging of words and correcting punctuation. Anne Self helped me with some such early editing and Tim Hyatt offered story suggestions.

My daughter, Melanie, has always been interested in my glasses missions work and was very helpful on a trip she made with me to Bogota. When I would write a story, I would have her read it. Melanie lives with me so it was easy to do. Sometimes the story got a thumbs up, sometimes a thumbs down. More often she gave me good suggestions for changes or additions. Thanks, Mel, for your valuable contribution to this effort.

I am so grateful to my son, Tim, who would have made a better Ranger than I ever was, for asking me to write and telling me about how I got started doing glasses work. Tim, a very important part of this book would be missing if not for you and your interest in what I was doing.

My brother Leon Hyatt, Jr. holds a Doctor of Divinity degree and has for many years written books on books of the bible. He first translates from the original Hebrew to English for study. Leon, of course, has a thorough knowledge of English and knows how to tweak and punctuate my long rambling sentences. I asked Leon to read some of my first stories. I wanted his thoughts about messages from God. He not only did that, but spent most of the day and half the night reading and editing and suggesting rewording some of my long rambling sentences. Leon, I want you to know I am forever grateful that God gave me an older brother to not only help me in this work but to give me guidance throughout my life.

All through the writing of these stories I have been continually plagued by unexplained problems. Complete stories disappeared from my computer, sometimes never to be found. Half stories have been found embedded in other stories. It was hard getting started to re-write the lost stories. I think they were never as well done as the original. Strange computer glitches occurred over and over again. They were all different, not the same ones over again. It was very frustrating and caused the work to take much longer than it should have. I think that Satan does not want this work to be published. I appreciate so much the work of my IT guy, Paul Guynes, who, dozens of times if not hundreds, over the last year has restored my computer, recovered some of the lost stories, repaired glitches in my computer, and replaced computers when they went out. I am on my third computer since I started writing. Paul has done all this without charge because he wanted to help me.

I don't have the words to adequately express my appreciation to my son Steve Hyatt who from start to finish has encouraged me and helped me with every aspect of this work; and as I write this, that help is ongoing. He has helped me know what stories I should write and which ones should not be published. He has tirelessly helped me with computer problems that constantly arose. He has arranged my random stories into an order that made them more readable. He has compiled an index and numbered the pages. Steve has probably spent as much time doing these things as I did in writing the stories. Steve, this work would not have been completed without you.

About the Author

Dick, Richard Dean Hyatt, is a retired General Contractor who resides in Lake Charles, LA in a home he built in 1965. He was a poor student and a high school drop-out who later further educated himself through reading and study. Hyatt considers himself to be an entrepreneur.

He is a member and regular attendee, once a leader, of Trinity Baptist Church, where both his parents and grandparents were charter members.

His Glasses Mission work has been the passion of his life since he first started it. The trips have been interrupted by COVID, but he plans to resume them in 2022.

Foreword

Friends and family have told me for years that I should write stories of some of the experiences I've had in life. Recently James Weeks told me again, and I decided to do it.

I selected some of the stories because they were truly incredible and some because the probability of them happening the way they did was astronomical. There are others that are merely interesting, if only to me. Some show me to have been foolhardy and dumb. Many show the passions that have driven me throughout my life. Together they record much of what I was, what I became and what I am now.

Writing these things has convinced me more than ever that God has given me these experiences, and that he has led me (when I would let Him) and protected me. There were times when I knew he was directing me and times when I was not sure.

Chapter 1: Early Days

Preschool

I have a lot of good memories of my preschool days. It was a wonderful time to be alive. In the summers, I wore only one piece of clothing, which my mother called coveralls. I wore pants and shirts and even shoes to go to church on Sundays. Dad had built a really tall swing set in our backyard. It had the usual swings and it had a bar to chin on. It even had a see-saw on one end. If dad hadn't built it so well, I would have worn it out.

Swing Set

There was a tall sycamore tree in the front yard in which I stored secret things high in the branches. I loved to climb way up in that tree almost to the very top. In those days, we didn't have hurricanes nor did we have good news reports about them. Of course, we knew they were hurricanes but we usually just called them storms; old men would look up at the sky say, "There's a storm in the Gulf." I loved to climb way up in that sycamore tree and ride it when the storms whipped it from side to side.

The winters were cold in the house, and we had only one heater. It was a little coal burning cast iron stove in the living room. I would grab up my clothes when I got out of bed in the morning and dress in front of the stove. We had a bin in the barn out back where we stored coal. We didn't have a water heater either. Mother would put the kettle on to heat up our bath water a little, but it didn't help much.

There was a privet hedge almost all the way around our backyard. There was a four foot high picket fence in one place where the hedge was missing. Someone told me that I would be a man when I could pee over that fence. I tried, but no matter how hard I strained, I never made it.

Dad had showed his three boys how to make pop-guns out of elderberry bush limbs and to shoot them with berries from the Chinaberry tree (neither of them grow here anymore). He showed us how to make kites out of split bamboo, newspapers and old rags. He taught us how to make slingshots out of tree limbs, inner-tubes (all tires had them then) and old shoe tongues. We shot rocks in them and got to be pretty accurate.

There was a little hole in the hedge. I loved to push my way through that little hole. When I did, I imagined that I was in another far-off land. There were huge oak trees there, and I imagined that the trees were full of bright-colored parrots and monkeys with long tails. Of course, I always had my slingshot with me when I explored the world on the other side of the hedge. When I crawled back through the hole in the hedge, I imagined that I was returning home from a far-off exotic journey, and everyone was happy to see me back after such a long trek.

One day, I went exploring on the other side of the hedge, fought my way into the big dense growth at the back, and came upon an old unpainted building. The only part of the building I could see had a big multi-paned window in it.

Well, there I was, standing in front of this big multi-paned window. Nobody had seen this old building in who knows how long. What would it hurt if I shot out just one of those window panes? Clinkity-clink. Man that felt good. I just had to shoot another one. A few minutes later, there were no panes left. Made one shot to each pane. No misses.

I went back through the hole in the hedge, feeling like I had conquered the world. But when I went into the barn the next day, I saw a big window with all the panes shot out. I felt awful. I had to go tell my dad. He just smiled and said, "That's okay, son." Dad was like that. That was during the

Great Depression. I don't think he ever had the window fixed.

Elementary School

When I was in the first grade of school, I was given glasses because I couldn't see the board well enough. I could see it, but not as sharply as I wanted. After a few years, my vision improved, and I didn't wear glasses again until I was pushing 50 and needed them for reading. Now after cataract surgery, I could do without them, but I still want to see everything as clearly as possible. I think that my desire to see clearly later made me want to help others to see clearly.

Miss Schindler

I wasn't a good student, but Miss Schindler taught me a lot. She was a famous English teacher in Lake Charles because she taught like no one else. A couple of generations of students who had her liked to brag about learning under Miss Schindler. There are not many of us left, but those of us who were taught by her will never forget her or the things she taught us. She taught us to speak publicly and to speak correctly. She not only taught English, she taught us things like table manners, etiquette and even kindness.

But Miss Schindler's crowning achievement was her "Literary Society." The society was not a requirement of her classes, and both the student and their parents had to voluntarily agree in writing for the student to become a member. She told us that if anyone did not want to join, it would be alright. She would assign some other classwork while we did society work in class. No one ever failed to join.

Her Literary Society teaching reached far beyond the classroom. Students monitored their classmates for correct speaking. We kept notes on fellow classmates speaking both in and out of school and recorded errors. Those errors were read aloud in class, and the student was fined a penny for each grammatical error. Vulgar words were fined two cents, and curse words cost a nickel.

A report of a curse word often brought a vehement protest by the one reported. "I said <u>Got</u> Damn, not God Damn." Protestors always got by with a two-cent fine.

I count myself as one of the very privileged to have learned under Miss Schindler.

Miss Schindler

Good Times in High School

One day when I was young, my grandfather gave me a very old Boy Scout manual. I really enjoyed reading about camping, cooking out, and living in the wilds. It was a very old version. The manual had been updated many times since mine was printed. I never got around to becoming really active in the Boy Scouts though. I was always too busy doing other things. Besides, I didn't need to join; I had the manual.

One of the things I found in the manual was that every scout should learn a poem. I wanted to learn "The wreck of the Hesperus", but it was too long. I settled on Alfred, Lord Tennyson's "Crossing the Bar". I can still recite it today.

Later in High School, there was a Scout troop called the Sea Scouts. The Sea Scouts had a sailboat named the Hi Can which was kept at anchor off-shore in Lake Charles. The guys would swim out to the boat and sail it around the lake.

I wanted to join the Sea Scouts but was told that I couldn't be a Sea Scout because I had never been a Boy Scout. Actually, I was a Boy Scout for a little while, but I let it go. I thought, "That's okay, I'll build my own boat." That was about the time Jimmy and I started building boats.

Well, one day, Jimmy and I were out in the middle of Lake Charles in a little skiff we had just built, when the Sea Scouts spotted us. They headed the Hi Can right at us and ran us over, swamping our little boat. Jimmy and I were in the water on either side of the Hi Can as it went by, so we each grabbed onto a shroud and swung aboard. It was all in good fun, and we had a good time sailing with our friends for quite a while before they dropped us off back at our little half-sunken skiff.

Take the Shot

Ike Hamilton and I rented a little boat and motor and brought our rifles along. We were just two young kids going nowhere in particular and looking for targets or whatever else we could find. I was in the stern steering the boat and we were traveling east just off the North bank of the Calcasieu River where it heads west below Prien Lake.

A buzzard was flying in the same direction off the south side of the river. Something in my head said, "Take the shot." I knew it would be a lob shot. I pointed my open sight .22 caliber semi-automatic rifle behind the buzzard, traced its flight pattern through the bird and out in front, trying to judge where the bird would be when the shot reached it and then still continuing the forward motion, I raised the sight way up trying to judge the trajectory of the bullet, and fired off five quick rounds. The shot connected after a couple of seconds, and the bird collapsed and crashed into the woods south of the river. A headshot. If it had been shot in the heart it would have stiffened, set its wings and sailed until it hit the ground dead with its wings still spread the way I have sense seen geese do. If it had been hit anywhere else, it would have fluttered, still trying to fly.

Ike couldn't believe what he had just seen. I could scarcely believe it myself, but of course, I acted like that was exactly what I expected. Looking on google maps today, I judge that it had to have been almost a 250 yard shot. What are the odds?

Anyway, that little experience helped me develop a can-do attitude: Give it a try. What have you got to lose? Ask for what you want. You won't do it if you don't try. Don't hold back. Yeah, take the shot.

Dropping Out of High School

I didn't like school work much and didn't make good grades, but the high school years were good years for me. They were the years that Jimmy and I fished and hunted together. We had a little pirogue tied out in the swamp that we used to catch bass in Contraband Bayou using long green slaughter poles cut from a cane break. We shot woodcock as they darted through treetops coming in to roost in the late evening. We shot wood ducks also as they came in to roost in little ponds along the bayou. We used headlights at night in the woods to shoot game and along bayous and in the swamps to catch bullfrogs. We shot squirrels in August when they were cutting pine in Lacassine woods. We always hunted barefoot whatever the season to be able to walk quietly in the woods. We set and ran trotlines across the Calcasieu River ship channel at the mouth of the bayou and caught loads of catfish. We used our little pirogue to run the trotline across the river to bait the hooks and to take the catfish off the hooks. Often tug boats came around the bend throwing a big wake, and they were almost on top of us before we knew it. That meant a fast paddle to shore before getting swamped. Sometimes we made it.

My brother Leon told me just recently that my mother had gone to my principal, Gervis W. Ford to talk to him about my poor grades. Mr. Ford said, "No, Dick doesn't make good grades, but he is smarter than a lot of the kids who do make good grades." Mr. Ford said just that. I have never been very smart.

Anyway, I went to Mr. Ford at the end of my sophomore year and told him I would not return. I was going to join the Army. World War II is what was going on in the world in early 1945, and I wanted to be a part of it. I was afraid it would be over before I could.

Mr. Ford said, "That's great, Dick; I'm proud of you. The Army needs good men like you." I thought, "Gosh, this is going to be easier than I thought." Then he said, "BUT, you have to get your high school diploma first. You will come back after the war and go on to college, but you will not come back to high school. The other kids will be too young for you."

When he saw that I was not going to give in, he said, "Do you know you have enough credits that in two months of summer school you could have your degree? All you have to do is work hard this summer, and if you can pass a test that Mr. Anderson will give you, you will have your diploma." If I had just thought about it, I would have known that I didn't have the credits. He was going to give me my diploma.

I breezed through summer school in a month and went to see Mr. Anderson, the Calcasieu Parish School Director. The test was to calculate the volume of his office. It was not simple. The office had several offsets. Some ceilings were flat. Some were slanted. Others were vaulted, and there was a bay window in the middle of one wall. I worked it out with a folding rule, pencil and paper. "That's close enough," he told me when I gave him my answer.

I think I did well on the test, but I think Mr. Ford and Mr. Anderson were working together and would have passed me in any case.

Chapter 2: Commercial Fishing with Jimmy

Oyster Lake

Oyster Bayou flows through the marsh and empties into the Calcasieu River south of St. John's Island in Big Lake. At the upper end of the Bayou, there used to be several small lakes. The uppermost and smallest one was Oyster Lake. I say the little lakes *used to be* there because now it is mostly open water.

One day, Jimmy and I put our net out in Oyster Lake and completely surrounded the little lake. We were just about to screw the net up and capture everything in the lake when all of a sudden, the whole lake erupted. I saw the tail of a giant redfish rise up at least five feet out of the water. My first thought was, "Whale." I knew it wasn't a whale. That's just what I thought as it rose up out of the water. There was no doubt that it was a redfish. We caught them all the time.

This one's tail had three spots. The largest was closest to the tail and almost the size of a volleyball. The next was somewhat smaller and about eight inches from the first. The third was also somewhat smaller. It was about half the size of the first and about eight inches from the middle spot. The width of the tail was the length of a normally real big redfish. I describe this fish in detail because if it ever gets caught, I want it known that it's "Jimmy's" redfish.

The fish was closest to Jimmy, and he was immediately on it. It was too big to be pocketed in the trammel net, so he was trying to wrap the net around the fish. Another huge redfish broke the water near me. It was just a slight bit smaller than the first one. So at this point, we are both wrestling our fish to try to wrap the net around them. Man, they were strong and, well, we couldn't hold them. They both got away from us.

The biggest redfish on record weighed 94 pounds. These fish had to be almost twice that. You can see pictures of the record redfish on Google. Nothing about it evokes the word, "Whale." Ours did.

If we could have landed those fish, our names would have gone down in fishermen's history books for as long as there were fishermen in the world.

White Lake

Jimmy and I had been fishing at Big Lake for some time, and we began to wonder if there were other good lakes to fish. So, we decided to check out White Lake.

Our trammel nets were made of linen. We took great pains to keep them clean of fish slime and never left them piled up wet. We always hung them up to let them dry; and we periodically tarred them. We had just tarred our nets when we headed out to White Lake.

A trammel net comprises of three layers of webbing hung on the same float line and lead line. The outer webbings have big webs, allowing the fish to pass through easily. The inner webbing has small webbing and is loose. When the fish hit the net, they push the small webbing through the big webbing, forming a pocket. They are trapped there until we take them out.

White Lake was amazing. The water was crystal clear, and the bottom sand was white as snow. Schools of fish were feeding all over the lake. We quickly put out our net, and surrounded a big school, but the fish had no trouble seeing the black net against the white sand in the clear water, and they would not hit it.

We tried hitting the water with oars, we tried screwing the net up tighter. When crowded, the fish would either jump over the net or find a way under it. All these fish, and we couldn't catch a single one. What a predicament! We just sat in the shallow water with the fish swimming all around us. After a while, the fish calmed down, and we discovered that we could reach out and touch them. Then we found out that we could stick a thumb in their gills and throw them into the boat. Wow! These fish were white drum, each weighing about 10 to 12 pounds.

When we go fishing, we bring a barge-type boat with big insulated boxes where we ice down our fish. At White Lake, we didn't have any ice because we were just exploring. We quickly filled all our boxes without even cleaning the fish. We didn't know if we could sell them or not. We took them to a little place on Lake Arthur that we knew about where they bought fish and they said, "Yes, we will take them just like they are."

When we first pulled up to the pier, one of the guys there asked, "Is one of you Dick Hyatt?" I was flabbergasted. I had never been there before. Then he told me my dad had called to tell me to come home; I had been drafted. I had told Dad where we would be going.

So, we unloaded as quickly as we could so we could get back to White Lake and more fish; but when we got there, the fish were gone. There wasn't a single fish in the lake. We put out our nets over and over and got nothing. No fish, no nothing. They were complete 'water hauls'.

I have told that story to fishermen around Lake Arthur and asked the guys where we sold the fish. Nobody had ever heard of such a thing before.

Birds and Alligators

One day I was by myself in a little john boat with a 7 ½ horsepower Mercury motor on Bayou Bois Connie when I decided to explore the little rice canal that cut through the marsh to the north. I followed the canal north for a while until it turned to the west. By this time, I was way back in the marsh.

What I saw on the west canal blew me away. There were low bushes on both sides of the canal. The bushes were next to each other and no more than about three feet tall.

Bird nests were everywhere. They were in the bushes – some holding as many as four or five nests – and on the surrounding marsh grass. Birds were everywhere. Some were in the nests on the eggs, some were feeding young, and some were sitting in the bushes and on the marsh grass. There were every kind of marsh bird that I knew and many more other kinds of birds. There was even a buzzard sitting on a single egg on a patch of bare ground. It went on for maybe a quarter mile. I stayed for a long time just watching all the activities.

Well, flash forward I don't know how many years, I am married with three young children. It is spring, and I want to show my young family the birds on the west canal. I loaded up Rissi and the kids and Melanie's peek-a-poo, Tuffy at Prien Lake Park in our boat, the 'Miss Rissi,' and headed to Bayou Bois Connie in Big Lake, towing a john boat behind us. We went up the bayou until we hit the North canal.

Tim and Dick Heading for West Canal

We all got in the john boat and started up the canal. There were big alligators in the canal, and every once in a while, we would hit one, and he would jump up, showering us with water. This caused Rissi to want to turn around and head back.

The water level was a little lower this time than it was before. We went through some places where we would drag the bottom for a short distance, and then we were in deeper water. I kept speeding through the shallow places, determined to get to the birds. Finally, we were hard aground. We couldn't budge.

I got out, turned the boat around and tried to pull it. I couldn't. I had to lighten the load. So, I told the kids to get out, I would walk ahead in the marsh next to the canal to scare the cotton-mouth moccasins away. Melanie was to walk behind me, Steve was to follow Melanie, and Tim would follow Steve. I told them to step where I stepped so the cotton-mouths would not get them.

That brought a howl from Rissi. She was beside herself.

"WHAT! YOU ARE GOING TO PUT MY BABIES OUT THERE IN THAT MARSH WITH ALL THOSE SNAKES AND ALLIGATORS?!"

But we all got out of the boat and when we did, Tuffy jumped out too and began to bark at an alligator. He had one backed up with its mouth open and Mel hollered "Dad, the alligator is going to eat my stupid dog." I put Tuffy in the boat with Rissi and she held him there. That gave her something to do. I waded the canal, pulled the boat and the kids walked along side as I told them.

Well, it all turned out okay, but we never got to the west canal; and I didn't get to show all the birds to my family. The birds may still go there in the spring. Someone let me know if you find them.

Big Pasture

One day, Jimmy and I arrived in Big Lake and began what was to be a three-day trip. In a short time, we had all of our iceboxes full. It would take a day to get the fish to market and a day to get back, and we wanted to catch more while the fishing was so good.

So, I came up with a plan. I would wade the marsh until I would get to high ground and then continue going east across the Big Pasture until I hit the Big Pasture Road. Then I would thumb a ride to Lake Charles and come back with a truck. Jimmy was to go from Bayou Bois Connie through the little rice canal where there was a small wharf. I would meet him there, we'd take the fish to market and be back for fishing that night.

The Big Pasture is now cut up with a lot of fences, but in those it was one BIG pasture. I started walking, and after a while, I could see nothing in any direction but open grassland. Not a tree or bush or fence was in sight. After a while, as I was walking east I saw some tiny spots on the horizon. They were north-east of me as I walked east. Later when I looked, I saw that they were cows grazing. A little later, they were all looking at me. Next time I looked, they were walking in my direction. I thought, "Surely they are not coming all the way over here." They were really a long way away. Soon, they were all running toward me.

What was I going to do? I grew up in town, I don't know anything about cows. The first thing I decided was I would not run. If I did, I would just die tired. I began to make plans. I would grab the first one, swing up onto his back and ride him as long as I could. No, I didn't think that would work! Then I began to wonder how long it would take for Jimmy to realize I was not coming back and how long it would take him to find me. I wondered that if I was not dead, how long I could live after the cows got through with me.

In the meantime, they were getting closer and closer. I thought, "If only I had a gun, maybe shooting a few rounds in the air would scare them away," or "I just had a stick, I could hit them on the nose." They were almost on me now, coming up from the north-west and I saw that there were about six or seven big bulls in front. They were HUGE. They were bucking and hooking and slinging snot. When they were about twenty feet in front of me, I looked on the ground for a stick, knowing there was none. I picked up a grass clod instead, threw it at them and yelled as loud as I could,

"GET OUT OF HERE!"

They all stopped. They were panting – big deep pants – and dripping snot. I had not broken my stride until I picked up the grass clod, and now I turned around and continued walking. After a while, I peeked over my shoulder and saw they were still standing there looking at me and still panting.

Later, they were walking around. Still later, they had gone back to grazing.

When I got almost to the big pasture road there was a Brahma bull in a really big pen surrounded by a very high chain-link fence. The bull took one look at me from the middle of the pen and charged right at me, stopping right at the fence. I thought those other bulls were big. This one made them look like runts! The bull then turned and ran as fast as he could south along the fence.

I thought, "Uh-oh, he knows where the gate is." I knew I could climb the fence, but he could reach me even at the top and I didn't want to jump down into the pen. I wasn't scared when the other bulls were running at me; I knew I was probably going to die, but I was not scared. But there was something about that Brahma running for the gate. I knew he would soon be out of the gate and on me, and that was after I thought I was safe. This time I was scared. I walked north and around the north end of the pen to the road without looking back.

I got back to town, picked up the truck, we got the fish to market and fished the next night.

Goose Hunt

There was a really huge body of geese off Oyster Bayou. Jimmy's brother, Bill was with us on this trip, and the three of us were fishing in the area. We met a man in another boat. After talking with him a while, he told us that the geese were on his land, and they were destroying it. They were eating all the vegetation, and it would become nothing but open water. There was no way to run the geese off. When he would walk or run at them, they would not leave; part of the body would just move out of gun range.

He asked for our help. He would furnish the shells if we would shoot them until they left. What a deal! We loved to hunt!

It took just one day to do it. We killed so many geese that now the problem was what to do with them. Jimmy said that his dad knew a man who bought illegal game. His dad would hang the game in a certain barn and call the man who would pick them up and pay him for them later. We just had to get them into the barn and call Jimmy's dad.

Taking them to Lake Charles in our boat would take too much time away from our fishing. There was always a game warden stationed on the road to Lake Charles where it crossed the Intracoastal Canal. He checked all hunters for licenses and counted their kill. Bill once got a big kill through the checkpoint by changing into clean clothes, putting on a tie and hauling his catch in a big travel case. We couldn't do that anyway because we had no vehicle.

So, I devised a plan. We had a small light skiff with us made of quarter-inch plywood, and we had a 10-horse outboard motor. I tied the geese together in bunches and tied weights to each bunch. If any boat headed toward me, the geese would go to the bottom of the ship channel. I covered the geese with a tarp and ran the boat, lying down on my stomach facing forward. I steered the boat by leaning side to side. The little boat literally flew across the water. I went up the little drain that goes through Brownsville, hung the geese in the barn, made the call, and was back in Big Lake in no time.

Okay, the deed was done, and the big body of geese was gone. But our adrenaline was still pumping and we wanted more. There were still some geese around, so we went hunting again the next day. Jimmy and Bill shot at one pond and I shot at another. I was only able to bag four, which was the limit.

While I was walking back and nearly to our boat, a single goose flew over me and I shot it. That gave me one over the limit. Then I noticed that there was another boat with ours. Uh-oh, game wardens. I threw my five geese down in the boat with the others; they had four each. The game warden told us that someone had been shooting a lot of geese and asked us if we knew who had been shooting there. We didn't. The warden checked my license and my gun.

I was lucky that he did not count the geese.

Rabbit Island

Rabbit Island is a little island in Big Lake. During the war dive-bombers used the island as a target. They dropped bombs that were filled with something white. The bombs would "explode" on impact and leave a big white spot where they hit.

When we were there, the island was full of old rusty bombs sticking out of the marsh and out of the shallow water around the island.

One day we were there knocking oysters off the bombs. We pulled our boat up to the island, and Jimmy walked off into the marsh. I was near the boat when I saw a huge cotton-mouth moccasin. It was the biggest I had ever seen.

The snake smelled the fish we had caught and was headed for the boat. I got an oar from the boat and put it on him so I could pick him up. I did that and was holding him behind his head in my right hand when another one almost as big headed toward the boat. I put the oar on that one with my left hand, he looked around and then at me. Then he started to slowly inch his way up the oar toward me. I couldn't keep him from making progress but before he got to me, I was able to get Jimmy over to give me a hand.

Neither one of us had ever seen such big cotton-mouths. We wanted to skin them, but we didn't have our skinning knives with us. They were back on the big boat several miles away.

So, I held the snakes, one in each hand while Jimmy operated the boat. The snakes were really strong and they kept wrapping around my arms and pulling their heads back into my fists. I kept unwrapping them from my arms and squeezing my fists as tight as I could, but I was losing the battle. By the time we got to the big boat, the snakes' heads were buried in my fists and were just about to pull all the way out.

We stuck needles in the brains of the snakes so as not to harm the skins. Then we skinned them out, heads and all. Jimmy brought one home and nailed it on his barn door where it stayed for many years. I brought mine home and put it on the wall in the den where we lived. My kids grew up with that snake skin on the wall, and it was there for many years after they were grown and gone.

A Roar and a High Note

One evening, Jimmy and I pulled the boat we slept in into the mouth of Bayou Bois Connie behind a large, tall canebrake to shelter us from a really stiff South wind.

Later, as we were preparing supper, we heard a loud, deep roaring sound from just outside the boat. There was also a high-pitched tone in the background. We had never heard such a thing before.

Running outside, we saw this huge black cloud undulating just out of reach between us and the cane. It was a cloud of mosquitoes. They got really bad in the marsh in those days.

Bus Ride to Morgan City

Jimmy and I took a bus from Lake Charles to Morgan City to look at a boat we were thinking of buying. The bus seemed like it was stopping at every wide spot in the road and was taking forever. I was getting tired of that and was getting impatient.

I told Jimmy I could hitch-hike faster than the bus was going, and so I pulled the cord to stop the bus out in the middle of nowhere and got off.

The driver asked Jimmy why I got off, since I had bought a ticket to Morgan City. He told him that I said I could hitch-hike faster than the bus was going. The driver said, "Oh he did, did he?" and the race was on.

In a few minutes, I caught a ride. As we passed the bus, I had the driver blow his horn, and I hung out the window and waved as we passed the bus.

A little later, I was on the side of the road again and the bus passed me, blasting the horn. Everyone on the bus was waving at me and enjoying the race.

That happened three or four times before we got to Morgan City. It was a lot of fun. Well, the bus beat me by about 10 minutes, but it turned a long boring ride into a memorable event for all of us.

Chapter 3: My First Love – The Sea

Maritime Service

Before I signed up for the Army, my older brother Leon came home from fighting in Germany. He begged me not to join the Army, and told me he thought I would like the Merchant Marine. I liked the idea.

Jimmy and I had planned to go to the Army together, but the Army wouldn't take him. Then the Navy turned him down as well, and then he tried the Merchant Marine; they would not take him either. Jimmy was strong as an ox, but he had a severe back injury when he was young, and that disqualified him. I signed on to the Maritime Service anyway. Would-be Merchant Seamen were sent to the Maritime Service to prepare for sea duty. It was much like the Navy's boot camp. The war in the Pacific was still going on; freighters and tankers were fitted with machine guns for protection from dive bombers.

They sent me to Catalina Island off the coast of Los Angeles to a training base. We were taught many things, like how to tie knots (I could already tie most of them) and box the compass. We also learned how to survive a tanker sinking into a sea of burning oil. We jumped into the cold water off of 30-foot high piers wearing cork-filled lifesavers that would hit us hard under the chin if we didn't pull down on them, and we swam underwater through tall seaweed.

Part of our training was also learning how to survive on a raft at sea even when we ran out of water. We learned how to do the work of the black gang (the guys below deck) in the engine room. We learned the work of the guys in the galley and we learned to be deck hands (my favorite), we learned it all.

They gave us hands-on training on an old ship that we sailed out into the Pacific. One time, we made port in Ensenada, Mexico; and had some fun shore leave there.

We wore a uniform similar to the Navy uniform and enjoyed some shore time in Los Angeles, mingling with the Navy guys and the California girls.

First Ship

I caught a tanker out of my hometown of Lake Charles and sailed her to ports in New England and back to ports on the Gulf a number of times. I got enough of that and wanted to sail overseas. In my ignorance, I got off the ship in Savannah, Georgia, thinking I could catch a ship from there across the Atlantic. I was wrong. Ships don't sail from Savannah to Europe and at that time it was very difficult to catch any ship out of Savannah. I was stuck there for quite a while.

I enjoyed walking the streets of Savannah and learned a lot about the city. There was a seaman's home on High

Battery called The Margarita that had been an upscale home. Sailors could rent a cot there for a dollar a night. There was a big pool in the front entrance of The Margarita. Drunken sailors quite often either fell in or were thrown in. After the war, the home was restored; it now remains as one of the tourist spots on High Battery.

Finally, I was running out of money. Of course, I knew that I could call home and Dad would send me what I needed, but I was not going to do that. With my last money I bought a sack of oranges and slept on a park bench. When I was down to my last orange, I was able to sign onto a coastwise ship as a messman. That wasn't what I wanted, but it wasn't bad. I liked the guys I was with, and enjoyed seeing how many orders I could take at one time without writing them down.

Then the war was over, and men no longer had to go through the Maritime Service to get their seaman's papers. When I heard that, I called Jimmy and told him to go to Port Arthur and get his seaman's papers. He did, I went home from New York, and we started sailing together.

Extraordinary Seaman

Jimmy and I sailed the SS Chicaca for 9 months. We took her across the Atlantic several times, visited ports in South America, Europe and the Red Sea. When I signed on, I was an ordinary seaman. I became a journeyman seaman while on that ship. I loved the sea. I loved the marine life, the whales, the flying fish, the dolphin, the killer whales, and the mackerel that would jump high out of the water.

I could go on and on. I loved the icebergs, the water spouts and the fire shooting volcanos that we passed. I loved the smell of the sea. I loved the storms at sea. I helped maintain the ship, stood watch on the bow and steered the ship. I worked 4 to 8, meaning I was on duty from 4:00 to 8:00 morning and evening. That also means I got to see the sunrise and sunset every day. Many were spectacular. I also got to see the breath-taking array of stars from the dark sea at night. It was a wonderful time for me.

One time we had a new first mate. Nobody liked him. We, the deck crew knew our job and we worked together like a well-oiled machine. When we would tie up at a dock, this new first mate would come up on the bow and shout orders, telling us what to do just before we started to do it. After a while I lost my patience, left the bow and went to tie up the spring line. That job requires at least three men; one man on the wench, one to take in the hawser and at least one to put on the stopper and belay the line.

While I was winching line in, Captain Smith called out "Let go your spring". I was getting ahead of the guys on the bow. I immediately let it go and began slowly taking it in again. He had seen me doing the work of three men and after that he always called me an 'Extraordinary Seaman' but that didn't mean he wouldn't jump on me if I did something wrong.

One evening, I was on bow watch. It was a blustery night. It was cloudy and the water was choppy. If the guy on bow watch sees a boat or other object or a light, he immediately reports it to the bridge. The weather continued to worsen as the night wore on. After a while, it got really rough and we were starting to take green seas over the bow. I was having a great time as the bow rose high in the air and plunged down into the sea. I was hanging on to the Suez Canal davit and enjoying every bit of it.

Then I saw a light two points off the starboard bow, and reported it. Captain Smith happened to be on the bridge, and he roared back at the top of his voice, "What in the HELL are you doing on the bow? Get your ass up on the flying bridge." Killjoy.

One day, we got a new 12-inch manila hawser for the bow. It would replace the sisal hawser we had been using. The sisal hawsers are stiff, they kink easily and are hard to work with. Manila hawsers are a dream. They are limber and work easily. Needless to say, the bow crew was happy with the new one. Soon after, while in Venice, Italy the manila hawser disappeared. It was discovered that the first mate had sold it. He was sent home and met by police when he got to the States. No one was sorry to see him go. We did miss our manila hawser though.

WWII Destruction

We sailed to European ports in Italy, France and Spain. We also sailed to England and even up the Thames River to London.

The destruction from German air raids was something to see. Boats and ships were sunk in almost every River or Bay we saw. Bridges were bombed out. Some buildings could only be identified by the parts left standing: Cathedrals and churches, schools and commercial buildings. Almost nothing was left of factories.

War is not a pretty thing.

Johnny

While we were off the coast of Leman, a bird flew onto the ship. It looked a little like a mockingbird, but somewhat bigger. He was not afraid of us but would not let us get close enough to touch him.

Someone called him Johnny and the name stuck. Johnny would walk around all over the ship. One day while I was in the fo'c'le Johnny hopped up into the doorway, looked around then hopped down and ran under a bed and came out with a roach in his beak.

That was Johnny's thing. He hunted roaches all over the ship. He became our mascot, 'Johnny the Roach Bird'.

Everyone talked to Johnny when he came into the room or when we passed him in the passageway. We all liked him, and we liked that he was catching roaches.

After a couple of months, while we were going through the Strait of Hormuz, Johnny flew away. He had rid the Chicaca of roaches and went on to another ship. It was sad to see him go. We all missed Johnny.

Fishing on the SS Chicaca

The Persian Gulf is a wonderful place. The sky seems to always have a reddish hue. Sometimes sandstorms onshore can be seen from far out at sea. There were many small fishing boats far out in the sea. We had to stay alert, especially at night, because they had no lights. The sea is teeming with fish. Numerous feeding schools can almost always be seen.

One day at a dock in Bahrain, there was a big school of fish all around the poop deck. Jimmy saw that, went to his fo'c's'le and brought out two lines with big silver hooks. We dropped the bare hooks down into the feeding fish and immediately started hauling in fish from about 30 feet below us. The fish were about two feet long with thick heavy bodies. In about an hour, there were so many fish on the poop deck that you could not easily walk through them. The cook and his helpers were busy cleaning and freezing the fish. Needless to say, we ate a lot of fish after that.

The Port at Caracas

As we were tying up the ship in Caracas, Venezuela, I saw a big shark swimming by the ship. Sometime later, the clear water looked so inviting I decided to dive overboard and swim to shore. I did, and some of the other guys followed. Somehow, I had managed to avoid the sea porcupines in the rough rocks when wading to shore, but all the other guys got the painful sticks.

After sitting on the shore a while, I got to thinking, "How am I going to get back to the ship?" If I walked to the dock, I would have to walk barefooted a long way across some very rough, sharp rocks. If I were to swim back to the Jacobs Ladder on the ship, how would I avoid the sea porcupines? I have never been very smart but I also got to thinking that maybe it wasn't so smart to swim where that shark was either.

The ship was tied up to a T-shaped pier and it was pretty far from where I was sitting. I was next to a bollard from where a hawser extended to the stern of the ship. The ship was light and riding high in the water. It was a long line and an uphill climb, but I decided I could climb it. I hung upside down by my hands and legs and began to climb. The first half of the climb went well; no problem. The next quarter was something else. The further I went, the steeper the rope became. The last quarter was sheer agony. When I only had about 10 feet left to go, I knew I was not going to make it. I could make it to the side of the ship, but not up and over the railing.

The next time I looked, a porthole near me was open, and I think the ship shifted in her moorings so that it got close enough for me to crawl through into an empty room. No one claimed to have opened the porthole or even being in the room. They all said, "I didn't do it." Not all T2 tankers had portholes, but the Chicaca had them. And this one was not only there, but now it was open and it was where I needed it.

Walking Fish

We were at a small dock way up the Euphrates River near Basra. The Euphrates forms the border between Iraq and Iran. After we tied up, most of the guys headed to town to visit the bars, but I walked up the Iran side of the river to see what I could see. I went through big palm tree farms where there were some round stone structures. I climbed up to look inside one and found it was half full of dates. A couple of times, I encountered young kids who were afraid of me. They would run away as soon as they saw me.

Later, I saw some fish in the river. There were places where the sandy land sloped down into the water. The fish would swim up to the land and use their pelvic fins to "walk" around on the land. Some would get 20 feet or more away from the water. They looked a little like catfish. I tried really hard to catch one, but they could run really fast back to the water.

Night settled in. The soft night sounds of the river, the birds coming in to roost and the animals calling out were different from the night sounds on a river at home. A long narrow boat with a dozen or more paddlers came down the river softly singing/chanting as they went by. I thought of the Old Testament prophet I had read about who was on a long journey and said to himself, "You are a long way from Jerusalem." I thought to myself, "Dick, you are a long way from Southwest Louisiana."

Solid Brass Shotgun Shell

When passing through the Suez Canal, ships are boarded by electricians who hang a huge arc lamp on the Suez Canal davit. All sea-going ships have the davits on their bows just for that purpose. A man gets inside the housing to operate the lamp, which shines a powerful beam ahead of the ship.

The electricians are joined by a guard who carries a 12-gauge shotgun. The shells he carries have full jacketed brass. They are beautiful things. One day, the guard was with several of us, just lounging in the galley when Jimmy asked to see one of his shells. After he looked at it, I took it, looked it over and passed it to someone else who wanted to see it.

When the guard asked Jimmy for the shell, he said he gave it to me… And you know the rest. The shell was gone, and I was responsible. The guard went ballistic. He broke down and cried, saying that he would be charged with stealing and his hand would be cut off. He begged me to take him to the captain. I had been to the captain's quarters several times with no problems, but this time he must have been in a bad mood. He threw us both out.

After an agonizingly long time not knowing what to do, Jimmy finally said, "Okay, here is your damn shell."

Joe Champlain

One day in port at Caracas, we had a sailing time of 5:00 pm. Our route out was to take us through multiple little islands. This day before sailing time, the entire deck crew and officers were at a bar. Captain Smith was there as well. I was there too, but I was not drinking. The captain was already feeling his oats.

Well, sailing time came and went; and Smith was still ordering more rounds. That went on for quite a while, and after dark, everyone staggered to the ship. Smith got us headed out and went straight to bed. There is always supposed to be one officer on the bridge of the ship, but that night they all sacked out. Everyone was out except for me and my friend Joe Champlain. I was steering, and my friend Joe Champlain was the lookout on the flying bridge.

Joe was my kind of guy. He was born on an island in the Caribbean and he had sailed square riggers. I think he spent several years before the mast when he was young.

There were a few commercial square riggers sailing at that time on routes between South America and Africa and occasionally to Europe. I was fortunate enough to pass one south of the Azores. She was cutting through the seas at top speed under full sail. What a beautiful sight. WHAAAT A BEAUUUTIFUL SIGHT. I could have watched her for hours.

Years before Joe had been in jail for a time in Genoa, Italy. If you have a little money, life is not so bad in jail there. Joe had a girl to stay with him in jail to cook for him and take care of him. While we were docked near San Remo, Joe wanted to go to Genoa to see his old girlfriend. At that time, toilet paper was not easy to obtain in Italy and was tightly controlled by the government. So, Joe stripped to his underwear and turned round and round while guys held toilet paper rolls that wrapped around his ample girth. Then Joe was off on a train to Genoa to smuggle toilet paper to his girlfriend, and I went along with him.

It was a beautiful ride along the seashore, through many tunnels cut through the rocky terrain. In Genoa, Joe went his way, and I spent some time seeing Genoa. On the train ride back, I was joined by a girl who spoke pretty good English. We had a great time laughing together and smooching through the tunnels.

Anyway, big old overweight Joe and I were communicating through the tube. He was not only drunk but said he felt really bad. Then he was having trouble breathing, so I tried to keep him awake. In a short time, he stopped talking.

So, I had the whole ship to myself. I was worried about Joe and wanted to go check on him. The wheel was slightly tight, and I thought it would hold its course while I ran up to check on him. He was really bad off. His skin was cold and clammy. I thought he was dying. I tried to think what I could do, but I knew I had to get back to the wheel.

When I did, we were *way* off course. In fact, we were more than 180 degrees off course. I had to complete the circle to get us back on course. Incredibly, nothing tragic happened and we sailed on. I was concerned about the tattletale that records the route we took, but nobody looked at it. Joe survived.

Captain Smith didn't get up the next morning. He didn't get up or regain consciousness for days. We all thought he was dying. We set up hourly bed checks around the clock, and on one check, found him with his covers kicked off, sitting up in bed reading a Western novel.

God had to be watching over me that night to keep us from hitting the rocks that surround those islands. To keep up from spilling out our 17,400 barrels of oil into the pristine blue waters of the Caribbean, and to keep Captain Smith, the mate on duty and me out of jail.

Orchids

Jimmy and I worked different shifts on the Chicaca, so we sometimes did not have shore time together. When we were in port at Caracas, I wanted to see the jungle. I found a guy with a jeep who worked for an oil company who agreed to take me.

The jungle was great. It was like any typical jungle, but I was having a ball. There were a lot of trees that they called Naked Negro trees. They were very tall and had no branches except way up at the top. The wet bark (it rains every afternoon in the jungle) was smooth and light brown and looked like – you guessed it – the skin of a Naked Negro.

But the thing that amazed me most was the orchids. They were everywhere. They were all over the trees, high up in the branches and also low to the ground. There were many different kinds. "Wow! What beauty." I began to gather them up. In no time, I had a jeep full of every kind of orchid you can imagine.

My mother dearly loved orchids. We had a distant relative named Onis Hyatt (everyone called him O. D.) who was head horticulturist at McNeese University and who raised orchids. O. D. knew about my mother's love of orchids, and he would often give one to her. She loved them and loved wearing them.

Well, I wanted to bring the orchids to my mother. I took them back to the ship and put them in a big shower room that was not being used. Every day, I gave the flowers a warm shower. We crossed the North Atlantic in the winter a couple of times, and we were in port in New York when they had a severe freeze, but my orchids were always safe and warm in the shower room.

Well, Jimmy and I wound up paying off of the Chicaca in New York in the summertime and went home. I put all the orchids in cardboard boxes and we rode busses back to Lake Charles. The bus drivers stowed the boxes in the luggage spaces below the bus.

Finally, we got home and I gave the orchids to my mother. She was delighted and put them on trees in the back yard, where she cared for them.

One day I got home and noticed that the orchids were gone. When I asked my mother about them, she told me she had noticed some small black specks on them and was afraid it could be some kind of South American blight. She had burned them.

Chapter 4: Moving On

As much as I loved the sea, I knew it was not to be my life. For the next several years, I did a number of things. I enjoyed and learned from them all. Here is a couple of them:

Traveling Salesman

For a while, I travelled for my dad, calling on wholesale grocers. Dad had a shoe store when I was very young. He went broke during the Depreciation because he sold on credit, and too many of his customers could not pay him. After that, he sold printing and learned that the bigger the order, the lower the price per unit. He wanted to find something he could sell a really lot of and decided on calendars. He signed contracts for huge numbers of calendars and sold them through wholesale grocers who sold them to the owners of Mom-and-Pop grocers that were everywhere before the days of the supermarkets. The wholesalers liked them because they were an easy sell to the stores, and they went for a huge mark-up compared to grocery items. The salesman liked them because they got to choose really nice products from a catalog according to the number of calendars sold. The Mom-and-Pop store owners liked them because their name was on the wall in the customer's home for a whole year. But someone had to go to the wholesaler and get them started.

That was my job. I covered almost half of the country. Except for a few cities in the far North East, I visited every city east of the Mississippi River, most of them twice. It was an interesting time. The cities were all different then. They looked different and smelled different, not like today because now they all look mostly alike, especially in shopping areas.

I stayed in cheap hotels, and on some weekends, I would find a safe place to leave my car and a safe place to put my cash. There were no credit cards then. I would put on old blue jeans and an old shirt, take just a few dollars, walk to the edge of town and thumb a ride. I wasn't trying to go anywhere. I just wanted to experience what it was like to be broke and alone in the world. How would I get along? Actually, quite well.

Once I was picked up by a man who had lost his leg in the war. The government had given him a car that he could operate with only one leg. We hit it off, and he took me home with him to spend the night. The next morning his wife put a whole loaf of bread in the oven to heat it. That hot bread and real butter were amazing! The next afternoon we were with a bunch of his buddies. They were all drinking beer, and after a while, they decided to get some moonshine. This was somewhere in Tennessee. Well, by the time they got the moonshine, they were already feeling no pain. I never drank on these outings; I didn't have the money, and I wanted to stay alert anyway. Since I was the only one not drinking, they got me to pour the moonshine out of a big jug into a smaller jug they could drink out of. I took a taste of the stuff, and it was really strong and hot. I spilled a small amount on my hand and even that burned. My friend continued to drive his car because nobody else knew how to drive it. That soon resulted in a fender bender. He didn't care. He said the government would fix it.

After another night and another loaf of hot bread, I was on my way back to my hotel.

Another time I was in Fayetteville, NC, and fell in with a bunch of airborne soldiers from the Eighty-Second Airborne. I really liked them and thought I would really enjoy being one of them. They didn't believe me when I told them I was not in the army.

Roughnecking

Working on a big barge oil-drilling rig in the waters of South Texas was an interesting experience for me. I liked working the floor and "rattling the iron," and particularly throwing the chain. I also loved to ride the block up to the monkey board high above the floor. It was a big rig, and the tower was tall.

I liked working with the men of the oil patch and was amazed at the strength of some of them. Since I knew something about boats and held a small boat operator's license, I spent some time operating the crew boat and spent nights on the boat tied up to the barge.

Chapter 5: Army

Becoming a Ranger

The Merchant Marine was an important part of the war effort, but it was not military. It was a draft-exempt job. Five years after WWII, the Korean War broke out, and I was drafted into the army and sent to basic training at Camp Polk, LA. I was a good bit older than most of the others and with a lot more of life's experiences. I looked at those little 18 year-olds that were drafted with me and thought that if I were to go to war, I wanted to have some more mature men around me. When the opportunity to volunteer to become a Ranger presented itself, I jumped on it. The 45th Division wanted 120 men to form a new Ranger Company that was to be commanded by Captain Charles E. Spragins.

Nine hundred men volunteered. After being tested for strength, mental attitude, etc. Spragins selected 300 men. Then he began a rigorous effort to make 180 men drop out. We were worked long and hard, night and day. I was in good shape from pulling trammel nets and roughnecking, and there was never a thought that I couldn't stay.

First, there was airborne training. They pushed us hard there, and we lost a few men either because they didn't want to work that hard, or they couldn't keep up, or they couldn't jump. They will "help" you out the door once, but if you can't jump on your own the second time, you are out.

We completed airborne training and got our wings, but we still had too many men when we started Ranger training.

Captain Spragins pushed us harder through Ranger training than we were pushed in Airborne training. At times we would train hard all day, slog through the Georgia boondocks all night, and train hard again the next day and into the night. If you can believe it, we learned to walk in our sleep. We would pick up little pieces of phosphorescence from the Georgia swampland and put them on the backpack of the man in front of us to keep us from straying. Finally, after more men dropped out, we were down to our 120 men.

We were trained in the use of every weapon that the army foot-soldier used at that time. We were taught to move at night and fight behind enemy lines. We were taught military tactics and survival techniques by combat officers, almost all of them the rank of Colonel and above. Then Captain Spragins taught us more tactics and had us train in their use. I especially remember his telling us to "Always close the back door," meaning send someone to guard your rear.

One day they read us a letter from the Commander of the 9th Ranger Company, which had graduated just before us and who were fighting in Korea. He told us that fighting is real and that we should learn well what we are being taught. He told us of men being shot when coming in for a parachute landing. He told us of men being bayoneted in their sleeping bags.

After hearing that, all the time I was in Korea, I slept on my back, holding a .45 automatic on my chest. I also bought a .38 special to strap to my leg for quick access on a combat jump.

Captain Spragins was a West Point graduate, and he set out to turn 10th Airborne Ranger Company into a lean, mean fighting machine of just 120 men. He quickly gained our respect because he never asked us to do anything that he did not do with us.

We were told that we were the best, and we turned out to be a cocky bunch but we learned well what we were taught.

When we were taught the use of explosives, some of the guys placed charges under the bridge to Phoenix City. A couple of guys got mad at Beechy Howard at the Phoenix City nightclub by his name and flushed a charge down one of the toilets. It was out of service for a while.

We were taught escape and evasion. Ranger Eddie Dunn, who was in my squad, was picked up one night for something and thrown in the stockade. Incidentally, Eddie was one of the two smallest guys in the company. The other was Joe Green who was almost cut in two by machine gun fire in Korea when his unit charged the enemy.

The next morning while we were in formation at revile, Spragins got a call from the stockade, telling him that they had Dunn locked up. Spragins looked out his office window at us and told them they should tighten up their security. Dunn had escaped and was standing in formation with us. Getting back to the base in the wee hours of the morning was not a problem for him because Spragins saw to it that all our guys had class A passes. We could roam the streets all night if we wanted to. That came in handy for me more than once.

Spragins taught us to look up and around when taking fire. See what is going on, decide what you can do about it, and take action.

We were taught how to quickly get off the streets and find friendly people in a hostile city. We were taught how to withstand torture. That has helped me manage and control pain throughout my life.

Many of the units that went through training at Fort Benning ran a course with full field gear and weapons. I think it was something like a 20-mile course. Spragins told us that we were going to run their course and that we were going to beat the all-time fastest record. We did it, and our record held for many, many years.

Captain Spragins made sure we had a lot of specialized training beyond normal Ranger training. He took us to Camp

Carson, Colorado to learn rock climbing and repelling. It was where the last of the horse cavalry was stationed.

When we first went to Colorado for mountain-climbing training, Spragins challenged the soldiers there, who were acclimated to the altitude, to a mountain run. In spite of the trickery that caused our guys to run further than the Calvary guys, the first man to cross the finish line was Ranger Bill Macaulay.

Dick and Rangers at Camp Carson Colorado

Bill in front row

We were taught to pack a mule at Camp Carson. Then we were taught rock climbing and repelling. We were dropped off high in the mountains, where our only duty was to survive for three days. We were given nothing but water. Emmitt Fike killed a snowshoe rabbit, and I was able to kill a bird with a big stick and roasted it over a fire. It looked like a fat little turkey and tasted good but needed salt. I also ate some berries and caught a salamander. I didn't get hungry enough to eat the salamander.

Then he took us to Seattle, Washington, and we were given the job of military police with the big MP badge on our arms. Man, the shoe was on the other foot now. We were breaking up fights and hauling drunken soldiers off to the stockade.

Captain Spragins was Scottish. He had dress kilts with all the trimming, which he sometimes wore on special occasions. On a side note, he got bagpipes for us, and several of the guys learned to play them. Our dress parades with the bagpipes were something to see and hear. We Rangers looked sharp with our jump boots, bloused pants, and the Ranger patch on our shoulders, but Spragins wanted us to look even more distinctive. He got 10th Company approved to wear the black beret as our standard uniform.

He ordered them for all of us. We got them just before we had a 10-day furlough and before we were to go to Japan.

I flew home wearing mine. I had a couple of beers on the plane and went to sleep. When I woke up, we were on the ground in Dallas, and people were deplaning. I got out and went straight to the latrine in a hurry. It was the first time I had seen the little men and women figures on the doors, and in my haste, I went into the ladies' room. A nice lady took me by the hand and led me out and to the men's room, saying very slowly and distinctly, "Not here; This is the men's room." She thought I was a soldier from another country and didn't speak English. I didn't say a word.

Anyway, we came to like the black beret and liked looking distinctive. Our little 10th Airborne Ranger Company guys were the sharpest in the whole army. Later the black beret became a standard issue for all Rangers.

Hokkaido Bear

When we first went to the island of Hokkaido, it was really hot. The heavy beer drinkers told me they could drink beer all evening and night and never have to go to the latrine. They would sweat it all out. I was never much of a drinker, but I did imbibe from time to time.

The symbol of Hokkaido is the black bear. There was a big standing bear in front of a little bar in Sapporo carved from the trunk of a tree. One night I was having a few beers with the guys at the little bar. When we left, one of them put his arms around the bear to see if he could lift it off the floor. He did. Then one after the other, everyone took his turn. I was next to last and got it up about 10 inches. Then Ranger "Bohunk" Nikiel picked it up, put it on his shoulder, walked about 50 yards, and threw it in the back of the 2 ½ ton truck that was there to take us back to camp.

When we got to the camp, Bohunk went off to bed and left the rest of us wondering what we were going to do with that bear. Someone had the bright idea of putting it in the latrine. The latrine was just a couple of rows of slit trenches surrounded by an 8-foot-high canvas wall. We stood the bear up inside the entrance.

The next morning while we were in formation, Capt. Spragins told us that a bear carving had been taken from the front of the bar and that our men had been accused. He asked if anyone knew about that. No one said a word, so Spragins reported that his company knew nothing about it.

Well, the bear stayed in the latrine for maybe a month. It was the soldier's latrine, so no officers ever went in. It became a common thing for us to salute the bear as we entered.

One day we had an inspection. Spragins inspected each one of us and our rifles individually. Then he went to our sleeping quarters and checked that out thoroughly. As we stood there in formation, he walked to our latrine and came out with a stern look on his face. He went to his company office and made a phone call. That day someone came and picked up the bear. We never heard any more about it.

Old T-10 Parachute

We have all seen pictures of people jumping from airplanes, sailing around on rectangular-shaped parachutes, and landing standing up on a very small target on the ground.

The T-10 parachutes we used were circular in shape, and we hit the ground at the same speed as if we had jumped off of a 14-foot-high platform, which we did in practice. We were taught to do PLFs (Parachute Landing Falls). You relax as you near the ground and hit toes first, twisting your knees in the direction of the fall and your torso in the other direction. Done correctly, it is a beautiful thing. You hit, roll, and you are back on your feet. There is very little you can do to steer the T-10 chute when you are in the air. You can pull in the risers on one side. That lets you slip to the side you are pulling, and it causes you to fall faster. After you are on the ground, there are at least a couple of ways to collapse your chute, which is often still open and blowing with the wind. The standard way is to pull the risers in and bundle the chute up in your arms. We were also taught that you could run around the chute until you are upwind of it, and it will collapse by itself.

One day we were making a demonstration jump for some group, and we were jumping without combat gear and backpacks. I chose that opportunity to try to steer the chute and also to run around it on the ground.

I did just that and found that you really can steer the chute a little bit. I hit the ground hard, and I also found it can be really hard to outrun the wind, but I was able to do it.

People have been known to survive falls from incredible heights using a well-executed PLF. People have also been killed coming in on a perfectly good T-10 chute. We had one of those in 10th Company. Ranger John "Jack" Daniels died on a jump in Japan. An over-zealous and thoughtless straight-leg officer had an ambulance driver park in the middle of the drop zone. Jack pulled up his knees as he swung over the vehicle and hit the ground on his tailbone.

Jack's funeral was very moving for me. Echo bugles were played while we stood in formation, saluting the flag. I was also saluting Jack.

Jack was the second of our men who were killed while we were in training. The first was Kennerd Duren while we were at Fort Benning.

Jumping is always a lot of fun for me. However, we were taught never to lose our fear because fear keeps us sharp and that the most important thing about jumping is mental alertness.

When making a training jump, if your chute does not open, you hit the ground in eight seconds. It should pop open hard from the prop blast in three seconds. If it does not, you have less than five seconds in which to react. You pull your reserve chute, which is strapped in front of your belt, and you help it to come out and open.

Sometimes your main chute opens but malfunctions, which causes you to come down faster in the air. There are things like a blown panel or multiple blown panels. There are cigarette rolls where the edge of the canopy is rolled up tight, making the canopy much smaller. One of our 10th company guys had five malfunctions on his first five qualifying jumps.

In the airplane, before making a jump, we are sitting in rows on either side of the plane with our chutes on. When we are about four minutes out, the leader calls," Stand Up and Hook Up." We do that, and each man checks the chute of the man in front of him, the last man in each row (stick) turns around to have his chute checked. Then the last man in each stick (if he is number 21) hits the man in front of him on the thigh and yells out "20 Okay" the next does the same and yells out "19 Okay" that goes on up the line until the #2 man hits the man in the door and calls out "2 Okay". The stick leader in the door then knows that everyone in his stick is hooked up, and all chutes have been checked.

The lead man has at least two things to help them determine when it is time to jump. He has studied topography and knows what it should look like when he jumps. There are two lights in front of the stick leader, one red and one green. When the pilot thinks it is time for the men to jump, he turns the red light off and the green light on; but it is the stick leader who is in charge. He decides when to lead his men out.

Most of the planes we jumped out of had two doors. The leader stood in one door, and the second in command stood in the other door. The number 2 man in the second door watches for the leader to jump. When he does, he hits the second stick leader on the thigh, and hollers go.

One day 10th company made a training jump in which Sgt. Bill Mathis broke his leg. He was laid up for months. When Bill returned to us, we were in Japan. We were making a training jump there, and Bill (being an NCO) was second in command and stood in the door. After not jumping for months and maybe being a little skittish about his leg, he was a little edgy. When the man behind him hit him on the leg and yelled, "Two okay," Bill jumped. He looked back at the plane and said, "I guess everybody else just froze." He came down through the thatched roof of a very surprised Japanese family and became the butt of 10th Company jokes forever.

The old T-10 chute is still in use, but the army is now in the process of upgrading its parachute. The new chute will be the T-11, and it is said to be the main chute and reserve chute in one. We'll see.

Hoss

Remember Hoss Cartwright? The big guy on Bonanza with the big hat? His real name was Dan Blocker.

Dan was with us when we were first issued our uniforms in basic training at Fort Polk. They didn't have the size 14 1/2 boots that he needed, so they told him to wear his dress shoes, and they would get him some boots. For the next few weeks, we were running through the hills of Fort Polk.

Every day we would go through the same routine. We all fall in first thing in the morning, and reports are made. Someone hollers, "All men present or accounted for!" But this day, the report was, "One man absent!" "Who is absent?" "Private Blocker." "Where is he?" "In his bunk."

So, all the brass go to the bunk, and Dan tells them, "The army promised me some boots, and I will fall in when I get my boots." His boots were flown in. He had them before the day was done.

We had two Rangers in 10th Company who had been cowboys together way down in Big Bend country of South Texas. One was Monte Goode, and the other was Bob

"Cowboy" Hill. They both grew up with Dan Blocker, who also was a cowboy.

One day Dan happened to be where we were, and Monte was able to get permission for him to fly with us on a jump. Dan enjoyed the flight and being with us, but he had no interest in jumping. You know, Dan could be a man of few words. He just said, "You guys are crazy."

Cowboy Hill was in my squad. One day on a jump after I hit the ground, I looked up and saw Cowboy coming in upside down. His boots were hung up in the risers. Cowboy was about to break his neck. I hollered to him to kick. He said he had been kicking all the way down, so I yelled at him to "kick different! Kick a different way!" Just before he hit the ground, his boot came free of the risers, and he hit feet first.

Another day we were making a night jump. Spragins gave us our instructions. It was to be a practice attack, and we would be slipping in on the enemy. He said everything would be quiet. There would be no talking in the air. All he wanted to hear was the quiet shush of the canopies opening.

When he jumped, Cowboy in his South Texas drawl hollered, "GEE RON A MOOO!!" at the top of his voice. The things Spragins had to put up with.

Going to Japan in Style

We have all known about soldiers being shipped overseas in the holds of ships, with hundreds of hammocks hung closely together. Spragins would have none of that. He managed to transport us to Japan on the SS General Buckner, which was more like a cruise ship. Not only that, but 10th Company's men were assigned huge state rooms, two men to a room, with big windows looking out over the deck and portholes looking out at sea.

Cup of Cold Water

One time in Hokkaido, I got really sick. I think it was some kind of Oriental bug or something. It hit me hard, and it hit me fast. I was hot with fever. I was thirsty. I went out back to throw up, and I passed out. When I woke up, my throat was parched; I was dying of thirst. I didn't have the strength to get up or even to crawl or to call for help. Nobody walked out where I was. After what seemed like forever, I got the attention of Homer Yowell and asked him for water. Homer would not have thought anything about one of us lying on the ground. We did a lot of that. He left and came back with a cup full. Homer was a big guy; he looked like a big angel standing over me with that water. What a relief. The water did the trick. Now I slept. I didn't pass out; I slept. It was late in the evening when I woke up. The sun was setting over by the latrine. That was when the Hokkaido bear was in the latrine. I was better now. The fever was broken. Jesus said that giving a cup of cool water to a thirsty man was like giving it to Him. (Dick's Liberal Translation)

Several 10th Company guys committed suicide later in life. One of them was Ranger Homer Yowell, a wheat farmer from Kansas.

I think that some of the guys who kill themselves are the ones who live their lives on their own terms and who take charge of how and when it ends. I also think that some of them don't see a solution to their problems and want it all to end. Maybe Homer was both.

I don't know what problems Homer was facing in his last days, but I wish I could have been there and had been able to relieve his hurting like he relieved mine on Hokkaido.

Dick in Hokkaido

Suba Party

All military in Japan participated in an exercise to "defend the country from attack". 10th Company was sent to the point where Hokkaido is closest to Russia. We could see the Russians firing off their artillery. They were having their own exercises at the same time.

Our duty was light there, and we had a lot of free time on the beach. A bunch of us went swimming one day and got caught in a strong undertow. We knew to swim parallel to the beach, but we were all in good shape, so we swam straight in. We could have saved ourselves a lot of hard swimming if we had done what we knew to do. Some of them barely made it.

One day some of the guys came to me and asked me to get Capt. Spragins to let me go into town to get some subas so we could have a suba party.

Subas are little watermelons, about the size of a really big cantaloupe. In Beppu, we used to get a lot of them and have suba parties in the evenings.

Well, I got permission and took Ranger Seals with me. The people of the village had never seen Americans before, and they were very interested in us. I had learned a little Japanese by then, so we could communicate a little, but not much. The kids followed us wherever we went. Seals and I went to a little store, bought a lot of candy, and gave it to them. We had a good time trying to communicate with the adults.

So now it is late at night, and we are in someone's house sipping sake and laughing it up with the Japanese when a jeep came to a screeching halt in front of the house. It was Sgt. Gower, who told me Spragins wanted me to report to him. "You mean he wants us to wake him up?" "No, he wants you to wake him up." "Man," I thought, "you are in big trouble now."

Well, he just asked me if Seals was back and if we were both ok. He never said anything to me about it, but I did notice that I seemed to get a lot of extra duty in the days that followed.

Train Ride to Beppu

During the Korean War, one Ranger Company was attached to a division of men. We were not well-liked. The "straight-legs" did not like us. We bloused out our pants at the top of our boots, making all the other's legs look straight by comparison. Some of the officers did not like us. They considered us to be wild and cocky and always getting into trouble. They were right about all those things. Rangers were given the most dangerous assignments and had the most casualties. Some of the officers complained that some of the best men were taken from their ranks and that they were hampered by their loss.

All that finally took its toll, and the Ranger Companies were disbanded. Spragins gave us the news. We were all stunned. It was a sad thing for all of us. Life as we knew it was about to change. We all would see combat, but we all regretted not being able to fight as a unit. We were given the choice of going back to the 45th division or to joining the 187th Regimental Combat Team. The 45th division was a straight-leg unit that was fighting in Korea. The 187th was an airborne unit stationed in Beppu, Japan. I chose to go to the 187th because I liked to jump, and I wanted to make a combat jump when we went to Korea.

We boarded a train for a three-day ride to Beppu. Talk about boring. We were trapped for three days with nothing to do but sit and listen to the same clickety-clack that all Japanese trains make. It's very distinctive from the sounds other trains make.

But finally, we were there. It was so good to be off the train. Wainwright and I reported to A Company. Captain Joy A. McDonald met us at the door in full field gear and told us to come in and make ourselves at home. They were going to make a jump. I loved to jump, and I was so eager to do something that I had this exchange with him.

"We'd like to go with you, sir."

"We are leaving right now, and you are not in field gear."

"We will be ready in two minutes."

So, guess what. We got on a train and went a full day to the north, got off, and trekked through rough terrain for three days before we flew back to Beppu and made the jump. Wainwright was not happy.

Transporting POWs

While we were in Beppu, Wainwright and I were often assigned the job of transporting prisoners - That's our American prisoners, not POWs.

At that time in Japan, there were several courier planes that constantly flew up and down the islands to every military facility in Japan and Korea, stopping just long enough to unload and load passengers. Any military man wanting to go anywhere would just go to the nearest airport, wait no more than 30 minutes and get on.

Wainwright and I transported all kinds of prisoners, from Japan and Korea - including those accused of murder. We wore our dress uniforms and carried loaded sidearms. We would fly to where the prisoner was held, pick him up and hop on the courier plane. Sometimes the prisoners would be in chains, but we would have them taken off. "Are you sure?" they would ask. We were sure. No prisoner was going to get away from two Airborne Rangers wearing .45s.

We would not ride the plane to our destination. We would go to the next stop, get off, take the prisoner to the stockade there and go and see that city. The next day we would do it again. It was a great way to turn a boring job into a great vacation trip. No one ever asked us why it took us so long. I have seen more of Japan than anyone I have ever met.

Kaio Maru

One day I was in the little city of Beppu, Japan, which overlooks a bay by the same name. On this day, I saw a beautiful square-rigger sail into the harbor and drop anchor.

A boat was off-loaded, and some uniformed men got in and went ashore. Just a short time later, the boat returned to the ship and was hauled back aboard. Then it hit me that they were about to hoist the anchor and sail away. I wanted desperately to board that ship, so I persuaded a girl named Yoshiko to go with me to translate, and we went down to the waterfront, found a boy with a boat, and got him to take us out. When we got to the ship, I saw her name on the bow. It was Kaio Maru. Yoshiko asked for permission to board, and we were invited on.

They not only allowed us to board but we were given a first-class tour. The ship and everything in it was brand new and spotless. The brass shone, the sails were white and spotless and furled perfectly; and the sailors were in full uniform and without flaw. They even showed us into the captain's quarters which were extremely plush. The hibachi was the most ornate and beautiful I had ever seen. Wow, what an experience!

Many years later, I learned that there was to be a tall ship sail-in in New York City and that tall ships would come from all over the world to be there. So naturally, I wanted to go and see my old ship, the Kaio Maru. Rissi and I watched from land while the ships sailed in, and the next day I went down to the dock and boarded the old ship. Old is right. To say that she had a lot of wear would be an understatement.

But I learned some things about the ship. The name Kaio Maru means World Circle. She had been a training ship for the Japanese Navy. During the war, her rigging was removed, and she was used for hauling coal. Then after the war, she was outfitted again. I had boarded her in Beppu Bay right after she had been re-rigged and returned to glory, probably on her maiden voyage.

Leadership School

Captain McDonald sent me off to leadership school. When the training was completed the leadership asked me to stay and be on their staff. It was not expected and I just told them "You would have to get Captain McDonald to agree."

Captain McDonald didn't say no, he said "Hell no."
I was flattered that they had asked but I would not have gone with them anyway.

10 Days in Tokyo

We were stationed at Beppu, Japan, with orders to be ready to ship out in 24 hours. My buddy, Leonard Wainwright, and I had a plan. We had a 7-day furlough coming to us, and we scheduled that along with a 3-day pass. We were required to check in between the two, but we had another of our buddies agreeing to check in for us after three days. Wow, we were going to have ten whole days in Tokyo!

We packed our things, put on our dress uniforms, and headed out. On the way to the gate, we decided to stop at the NCO Club and have a leisurely breakfast. When we left the club, we saw vehicles speeding around and men running everywhere. Just then, a Colonel pulled up in a jeep and told us to return to our unit. We pleaded furlough, but that went nowhere. If only we had skipped the breakfast, we would have been gone. We were on our way to Korea in just 14 hours.

T-bone was our company cook. He earned that nickname when Spragins had T-bone steaks delivered to us and little later asked him when we were going to have them. T-bone said, "Is that what those were? I made stew out of them. We had it yesterday."

Anyway, in his haste, when we went to Koea, T-bone failed to pack salt, so we had no salt for a good while. T-bone used a lot of bacon in his cooking during that time. His bacon biscuits were especially good. Later back at home, I made bacon biscuits for my family many times.

Koje-Do

Koje-do (do is the Korean word for "island") is a small island south of South Korea. During the war, our POWs were sent to Koje-do, where they were kept in big barbed wire compounds. We were rushed in because the POWs had captured the American general and some of his aides.

Gen. Dodd and his aides were released before we got there, and our job for several weeks was to clean up the mess.

Koje-do was a mess. At that time, there were 80,000 prisoners on the island. There was a hierarchy in the prison system. There was the hard-core career North Korean soldiers. There were the rank-and-file North Korean soldiers, and then there were Chinese who had been sent to assist the North Koreans in their fight. The hard-core guys ruled the prison, the rank-and-file followed, and the Chinese hated the North Koreans and wanted to escape.

Their latrines consisted of 55-gallon drums cut in half with wooden structures built over them. The drums (honey buckets) were fitted with bales for carrying. The prisoners periodically carried them down to the bay and dumped them. Flies were everywhere. They would get in your nostrils when you breathed. They covered your food when you tried to eat. The stench was everywhere, but we soon got used to all that.

The fences were 12-feet high, topped with 4-foot-long slanted pieces. The barbed wire barbs were extra-long and extra-sharp and close together. The strands of barbed wire were 6 inches apart horizontally, with vertical pieces woven 6 inches apart. There were two such fences about 30 feet apart, with big rolls of barbed wire entanglement in between.

Most of the hard-core guys were in compound 76. They were the ones that captured Gen. Dodd. There was a scattering of the other two types throughout all the compounds. There were many signs on the compounds reading, "GI if you value your life, do not come into compound 76", "Death to America," etc. Regular military training was held inside the compound, mostly bayonet training. There was a lot of loud yelling and chanting.

The compounds were, of course, guarded by the US military on the outside but also by North Korean guards with bayonets posted every 30 feet on the inside.

Sometimes a Chinese POW would attempt an escape. It was a daring move. He would have to avoid being caught and bayoneted on the spot by a North Korean guard. A good many of them were buried in compound 76. Then he would cut himself up climbing the fence getting through the barbed-wire entanglement and the next fence. Then he faced the danger of being shot by a trigger-happy G.I. who did not understand what was happening. Those that made it all the way through would immediately take off their cap with the North Korean insignia, throw it down, then squat and put their hands on their head.

Each compound had its own kitchen where the prisoners cooked their own food. Compound 76 also had a carpentry shop where they fashioned wooden guns and a foundry where they made bayonets from 55 gallon drums and other metal.

We began our task by building more compounds so we could separate the different types of prisoners. After we built a new compound, some of the guys had races to see who could cross the barrier the fastest. I declined that one.

Then the big day came when we were to enter the infamous Compound 76. The prisoners knew we were coming. They had read it in the Stars and Stripes, which they picked up from the latrines. They had their bayonets. They had Molotov cocktails made from their cooking fuel. They had grenades made from shells picked up by the honey bucket brigade when they cleaned our latrines, and they had some wicked-looking flays made from sticks and barbed wire.

Our men had unloaded M-1 rifles with fixed bayonets, and NCOs had concussion grenades and loaded .45s. First, the NCOs were directed to toss the grenades over the fence at the inside guards, who immediately retreated from the fence. Many of the guards were blown up. Later I saw a picture of a pile of arms and legs that had been cut off after being shattered.

A tank was pulled up to the front gate, and while everyone's attention was focused there, we went into action in the back. A team of engineers quickly cut through the outer fence, dropped a boardwalk over the barbed-wire entanglement, and cut through the second fence. Captain Joy A. McDonald was the first to walk through, followed by, you guessed it, my buddy Leonard Wainwright and me. We were to form a line at the fence and begin to move forward to clear the buildings. An engineer with a flame-thrower was to come behind us and burn down the buildings after we had them cleared. But the flame-thrower guy behind us got ahead of his game and moved in with the line. He was between Wainwright and me, and the first thing that happened was one of the gooks tossed a grenade at the flame thrower. A medic was on him immediately, and as he worked on him, the medic was hit by a Molotov cocktail and was on fire without even noticing it. I started hitting him to put out the fire, and he, an officer, looked at me incredulously. I said, "You're on fire, sir," and he turned back to his patient, who had been hit in the groin and later died.

Wainwright and I were both hit by the shrapnel. I moved my legs and wiggled my toes, and everything worked so I didn't slow down.

We started going through the bunkhouses, driving the prisoners out - the men with their bayonets and me with my .45. The men encouraged the prisoners to move along by bayonet jabs to the butt, and I by raps on the head with my .45. One of our men came to me and said, "Sergeant Hyatt,

I've been hit." He had been hit by the shrapnel that got Wainwright and me. I said, "Go back and find a medic - but give me your rifle." I had seen that prodding with a bayonet was more effective than tapping on the head with a .45.

Later, Captain McDonald saw that I had been hit and told me to go back. I told him I was alright but got a stern, "That's an order, soldier." He had just shot and killed a prisoner who had just bayonetted one of our men.

So, Wainwright and I got some shrapnel cut out of our legs in a MASH unit. Afterward, we lay on our bunks side by side with our feet propped up on something at the foot of the cots. Our legs had been completely shaved, and they were the best-looking legs we had seen in a long time.

I had a deep incision clear across my left calf, and the Doc told me not to walk on it for at least two weeks. We were then sent to a facility where stairs were pointed out to us, and we were told we would be on the third floor. So much for "Stay off your feet!"

We were told that we could stay there as long as we wanted. They said nobody would ever tell us we had to leave, so we thought, "This is great. We only have a few months left; we will spend it right here." We soon found out, though, that there was nothing there, nowhere to go, nothing to do. One day a train came in carrying a big load of wounded soldiers. Many of them were laid out in the halls because they didn't have enough beds. Wainwright and I looked at each other and said, "We're out of here." So, we were back walking guard duty between the compounds before the Doc's two weeks were up.

Years later, I saw my V. A. doctor for a regular visit. She ran some tests. That night she called me and said the vein in my leg showed it had been severed. She said your leg was warm, so I know there is circulation. She said the blood had found a new path through the calf muscle in my leg and developed its own vein. She said that had to have been excruciatingly painful for a very long time. Well, you know, Rangers are tough.

Sam's Patrol

Sam Payne was a sharp soldier. He was always neat. In Japan, there were times that the sharpest soldier would be recognized and given some small award. Sam always won it.

Our winter dress uniforms were wool and would hold a crease for about 1 minute. But Sam's creases would last through rain or snow. How could he do that? He wouldn't tell me. We were in Japan. A Japanese boy named Akio lived in our barracks and took care of our laundry, and pressed our uniforms. Sam would always tell me, "Akio presses my uniforms just like he does yours."

Finally, after a lot of pestering, he told me. One day he had spilled beer on his uniform, and when it was pressed, the creases stayed in where the beer had been. After that, he always put a little beer on his uniforms before he had them pressed.

Later we were in Korea. At the time, we were dug in, and the Koreans were dug in out in front of us. We would go out in platoon-sized patrols and set up ambushes. The Koreans would do the same. We never made contact.

Sam and I were both squad leaders in the same platoon, and one day, he was given a different kind of patrol. He was to take his squad out later than usual and was to go farther than usual. He was to find a certain checkpoint before returning.

I wondered why my squad wasn't given the job. I was the only Ranger squad leader in the Company. If there was to be a difficult job, I should have been chosen for it. Since that didn't happen, I wanted to go with Sam.

I used the BAR (Browning Automatic Rifle) as a reason why he should take me. There was one BARman to a squad, and Sam had lost his. I told him he was likely to see some action, and he might need the extra firepower. I would go with him and carry a BAR. I liked the BAR. I was a BARman for a while in training. He thanked me. He said he didn't feel good about the patrol, and he would like for me to go.

We were to meet at 3:00 the next morning. When I went out to meet him, he was already there with a few of his men. He saw me coming, left his men, and came over to me. He had changed his mind and said that I could not go. He said he didn't want his men to think he couldn't handle the job without a Ranger coming along. No amount of pleading would change his mind.

Of course, daylight caught him right in front of the enemy. From where I was, I could see them coming by the thousands out of their underground bunkers. It looked like swarms of ants coming out of an anthill. Sam told his men, "There'll be no prisoners."

I saw a lot of bravery that day. A lieutenant with a platoon of men came toward me at a fast pace. He asked me where the path was through the minefield. I told him he had missed it and I would take him to it. No, he said, we are late already. We are going through right here. He told his men to follow him, staying two steps apart. He told them to step where he stepped, and when he went down, the man behind him would be in charge. He also told them when that man goes down, the next will be in charge.

I saw a sergeant ordered to take his platoon out to the left flank to draw fire. He took off his helmet, tied a white handkerchief around his head, and ran out yelling, "Hey you Sons of Bitches, Shoot at Me!"

Meanwhile, Sam and his squad members had all been hit and were down in between the rocks, picking off anyone who came in sight.

Sam's radioman was Cpl. Lester Hammond. The radio was a heavy backpack thing. Hammond was a pudgy little guy whose dad had a lot of pull in Washington. His dad didn't want him to go to the Army and had the clout to keep him out, but Hammond wanted to go and even went airborne.

Hammond crawled up on the top of a rock and called in artillery. He was directing fire where he could see the most enemy. Sam and the others kept telling him to come down, but he wouldn't do it. He had been hit with others and then was hit a couple of times more. As the gooks, as we called them, got closer and closer, he finally called artillery in on himself. Cpl. Hammond was credited with killing 250 some-odd enemy soldiers and was awarded the Congressional Medal of Honor. It took us all day to get the guys out.

As I said, all of the men in Sam's squad were hit, and we never saw any of them again, except Joe Damato, who returned to us slightly wounded. Joe told us what had happened. He also told me they didn't need a BAR. Their M1s did the job just fine.

On a side note: Why were Sam and his men sent on that patrol? Why was Hammond sent to his death? Why were they sent to a checkpoint that probably did not exist? Why others were killed and wounded that day? Well, I figured it out.

The top career military men were not getting enough action. They needed some real combat experience on their records to help them advance in rank and prestige. The operation was all pre-planned. That's why they were able to respond so quickly and so well. They got what they wanted, and they got it in spades. It was a big event, a major enemy kill. And by their quick and decisive action, there were not too many of our guys killed or wounded, and (the icing on the cake) there was a Medal of Honor winner. I surmised that and later told Pete Spragins my suspicions, and he told me, "That's exactly why they were sent out."

Wonder What the Gooks are Doing

While we were in that static position, we fired 50-caliber machine guns at the enemy positions; and we both fired artillery rounds at each other. Incoming fire was a common thing. A round or two would come in every once in a while.

Incoming fire sounds very different than outgoing fire. It was also easy to tell how close rounds would come to us. When the rounds were to be close, we would bend over a little bit. When they were closer, we would get down.

We had a cook that wanted to be relieved from cooking and allowed to join us on the line. He was given a bunker next to mine. He went in for the first time, and right away, a round hit at right at the entrance to the bunker. He wondered for a few minutes if he had made a good move.

At one time, a new guy in the outfit was handling a .45 for the first time. I was standing close to him, and others were all around us. The gun accidentally went off, and it really shook the guy up. He tensed up, and it went off again and again. He was turning around and around, and lead was going in every direction. I shouted for him to throw it down, and when he didn't, I rushed in and took it from him.

It surprised me that whiskey was readily available in that situation. I can't remember the brand that everyone seemed to have, but I remember it cost $11 a bottle.

One evening, four Master Sergeants from four different companies met together to have a few drinks. After some time of drinking and feeling no pain, they wondered what the gooks were doing over there and decided to take them a bottle of whiskey.

So, the four of them got in a Jeep flying a white flag and headed out to see what they were doing. It took us quite a while, but we found the burned-out Jeep and recovered the badly-mutilated bodies.

Bouncing Betty

While we were online in Korea, Chinese soldiers would sometimes slip away from the Korean line and cross no man's land. At daylight, they would be standing in front of our line with their hands on their heads. We would escort them through the minefield and turn them in as prisoners of war.

It happened often enough that the brass wanted another pass through the minefield. Sgt. Richard Childs was given the job. He took a few men with him. Clearing a minefield is a slow job. You push a bayonet into the dirt at a slant to detect anything under the ground. That has to be done with every inch of ground you are clearing. Sometimes you encounter a string that runs off to the side, and you have to continue the procedure along the string until you get to whatever is at the end.

Childs and his men started the job early in the morning, and at about 11:00, they hit a string. Before noon they got to the end of the string, found it to be a flare, and took a lunch break. By now, it was really hot. Business picked up then, and by 3:00, they had found three more flares. And at 3:30, another string. It was hotter than ever, and the men were exhausted. Sgt. Childs said, "Another damn flare," and reached down and pulled the string. It was a mine - A Bouncing Betty! It jumps up and explodes above ground.

This mine is not meant to kill; it's meant to injure. A dead man takes one man out of the fight. An injured man takes three men out of the fight, the one that is injured and two to take him back.

One of the men laughed and hollered out that he was "hit in the prick and can't f*** no more." This is the way many react to injuries. They don't scream like in the movies - at least not our kind of men.

Childs and his men were back with us in a month or so.

Dick in Korea

Leaving Korea

The day came when I had completed my tour of duty. It was time to leave Korea. It was a sad time for me. The job was not finished. I didn't like leaving my men. I felt a little like I was a deserter. But it was time; my tour had come to an end. After goodbyes and promises that we all knew would not be kept, I left. There were six of us that day going back. We were told to wait a little while; the Col. wanted to speak to us.

That would be Col. William Westmoreland, who later was General Westmoreland, who commanded the Korean War and later was commander of all Pacific operations.

He first commended each of us. He said that we had done a good service for our country. Then said we should not expect anything like the welcoming parades and praises our older brothers and uncles had received when they returned home after WWII. He said he had just returned from the States, and the mood there was very anti-war. He said that we could encounter hostility because of our service. He again thanked us for what we had done, shook our hands, wished us the best, and he was gone.

I have always been amazed that Col. Westmoreland would take time out from running a war to speak to six ordinary foot soldiers.

Master Sergeant

After my tour of duty in Korea, I was sent to Fort Polk to be discharged. There was a Yankee Master Sergeant there overseeing the filling-out of forms by a room full of guys being discharged.

The sergeant started off denigrating the South and the dumb yokels that lived here. He got off on blacks, and a young black man objected. That led to an even greater tirade. He used the analogy of knives for words and cut the guy down with his words. He said his knives were sharp, and the guy would not want to exchange words with him. The man was shut down.

Then he laid down the rules. No one would speak unless spoken to. We would not turn around to our tablets unless told to do so. We would not return back to face him unless told to do so.

He went on about how to fill out the questionnaire before us. He said he had to explain it in very simple terms so that we could understand it. We were all fuming. I couldn't do anything because the sergeant had one more stripe than I did and I didn't want to cross swords with him anyway. So, we all complied.

After a while, to my delight, he slipped up. We were turned around facing our tables when his tone changed, and he began to talk to us in a normal tone. Everyone turned around to face him without being told as he went on talking. Everyone that is, except me. He knew he was caught. I winked at the guys facing me as he went on talking as if nothing had gone wrong. They were all too scared to acknowledge my wink in any way. After a while, he said, "Now, everyone, look up here." and I turned around.

Later, we had turned in the form and were working on another one when a Corporal came to me and said that he knew that French was spoken in Louisiana, but we should use English words on our questionnaire. Now I knew the sergeant was afraid of me. He had sent a Corporal to do the job for him. For the question, "What do you plan to do or to be in civilian life," I had just written the word "Entrepreneur."

"Corporal," I said, "You will find that word in the English dictionary, or you can just ask the sergeant there. He'll tell you what it means."

Ha! Gotcha, Sarge!

Welcome Back To Louisiana

About 20 years after I was out of service, I read that General Charles Spragins was to be the next commander at Fort Polk. Could that be Captain Spragins that I had known? It had to be. I sat down and wrote him a letter, welcoming him back to Louisiana, and addressed it to him at Fort Polk.

He told me later that when he was shown into his new office, there was nothing on his desk but my letter.

One of the first things he did was to contact 10th Co. Ranger Billy Sinor, who had stayed in the Army and followed him throughout his career. Everywhere that Spragins went, Sinor followed. On this move, the only way Sinor could follow him was to take a job as chaplain's assistant, a very unlikely job for Sinor. Spragins told the chaplain, "Either Sinor goes to heaven, or you go to jail."

Spragins gave Sinor his first job at Fort Polk. He told him to round up all the old 10th Company guys that he could find. He said we were going to have a reunion. Sinor did a good job, and a lot of us got together in Tulsa, Oklahoma. Sinor and a lot of our guys were from Oklahoma.

We had reunions every year after that, each hosted by a different Ranger. After a few years, we started inviting wives and other family members. The wives thought every year was too often and got us to make it every other year. We continued to meet for reunions until a few years ago. A few of us still meet for special gatherings and funerals.

After we were out of the service, our old Company Captain was Major General Charles Spragins, but he insisted we call him by his nickname, "Pete."

Pete called for the first 10th company reunion, hosted a couple, and attended almost everyone. He inspired us when he spoke to our group. He attended our funerals and regretted when he could not.

Here are a few excerpts from letters he wrote to me through the years, all written by hand with a fountain pen:

"What a truly singular and stand-out experience that brief but fast-paced ten months we served as 10th Co. Rangers. NOTHING, before or since in my military experience compared to it."

"The most enduring dimension is the lasting friendship and genuine comradeship - it's an absolutely UNIQUE experience – and I truly treasure the memories and friendships that have bound us together."

"Dick, you are a very large part of the glue that has kept us together with your faithfulness through this."

For many years Pete had said I was the glue, and I have told him that he was the glue. He was because of who he was and what he has meant to us. He called me the glue because I had written a newsletter that went to all 10th Company men and later many of their widows. The letters were regularly published in the quarterly newspaper The Ranger that went to all US Army Rangers. I still write the newsletter; it goes to about twenty 10th Company Rangers, a good many widows and others.

After Pete died, his wife Cena sent me Pete's favorite dress jacket, which I had altered to fit me. It has a small

Ranger tab pin on the lapel and an insignia on the pocket, which has a Ranger tab on the top with a skull and cross rifles under it and with the words, "Mess with the Best, Die like the Rest." He often wore it to church. I haven't done that yet, but I may someday.

General Charles "Pete" Spragins

Dick in Pete's Jacket

Leroy Jerry Carl and Dick

Emmitt Luby and Dick

Carl Johnson

Carl, Emmitt, Pete, Fick Guest, Art, Ralph, Marino, and Dick

Bill, Dick, Ralph, Ames, Carl

Carl, Art, Jerry, Baxter, Dick, Eddie, Marino and, Brittin,

Emmitt, Ames

Award Knife

More on the Black Beret

Well, it turns out that General Shinseki, who commanded an artillery brigade, had wanted to have his brigade approved to wear the black beret; but he couldn't get it approved. Imagine! General Shinseki couldn't get his unit approved, but our Captain Spragins got it done for a single company. Unbelievable!

Much later, it became a standard issue for all Rangers. Still, later Shinseki had risen to the rank of Lieutenant General (three stars); and now he had the clout he needed. He made the black beret the standard uniform for all U. S. Army soldiers. Rangers everywhere were furious. It started a huge fight within the military that went on for quite a while, but the dastardly Shinseki won out. Rangers were given a choice between continuing to wear the black beret or wearing a tan one. They chose the tan beret but didn't like it one bit. 10th Company continued to wear the Black Beret at our reunions.

It was soon learned by Congress that Shinseki had bought black berets for the entire army from a firm in China. That caused a huge stink that ended with Congress demanding that the berets not be issued and that Shinseki buy them again from firms in the U.S. The problem was that there was no firm in the U.S capable of making the berets, so they had to learn how and had to buy the machines that could make them.

So, Shinseki bought $10,000,000 worth of black berets from suppliers in China and then had to buy over $15,000,000 worth of the more expensive berets from suppliers in the US and other Allied countries. Google quotes much smaller numbers from a Chicago Tribune article, but common sense shows them to be ridiculously understated. Anyway, whatever the numbers, Shinseki lost a lot of the taxpayer's money and, worse than that, he lost face, a terrible thing for a Japanese man, and Rangers got a little bit of satisfaction.

Carl Johnson's Reunion

Carl hosted a reunion, and he held it in South Korea. Carl had gone one step further and made available to us one of two side trips to combine with the Korean trip. One was a several-day stopover in Hawaii, and the other was a tour of China. Carl and his girlfriend, Cherié, Freddie Dyer, and his wife, Ruth, and I were the only ones that chose to go to China. After the Hawaii goers heard our stories, they wished they had gone to China too.

The tour of South Korea was made available to veterans who had fought to save their country from being overthrown.

The South Korean government even paid for much of the cost.

Korean Leg

Modern Korea was a revelation to me. Gone were the people with the traditional Korean dress. Gone were the primitive tools of the past generation. Gone, of course, was the destruction from the war. Instead, we saw all modern buildings; people dressed much as we do. We saw industry and commerce. One night from a high point in Seoul, I saw hundreds and hundreds of neon cross signs all across the city. After the war, hundreds of Christians from the United States went to Korea to tell the people about Christ. Christianity took hold, and now almost 30% of the people are Christians.

We went to Inchon and visited Freedom Park and Inchon Memorial Hall. I was impressed by the monument of "Tall spires separated by an ocean reaching toward a common lofty goal, and huge broken links below, which are reminders of the connection between two countries whose ties are broken, but whose heritage is forever entwined."

We toured the Korean Folk Village, which depicts buildings, people, and customs of old Korea before the war. For us, it was a look black to Korea we knew during the war. No one in South Korea lives that way today.

We visited the DMZ at Panmunjom and the small building where peace talks were held. There was a stripe through the middle of the building and the surrounding land showing the border between the two countries. We were warned to be careful not to step over the line and not to speak or engage the North Korean soldiers on the other side. At one point, I was close to a North Korean soldier, and I thought to myself, "The last time I was this close to a North Korean soldier, I stuck a bayonet in him." The last year of the war was fought while the final sticking point was hammered out in that little building. The issue was one of repatriation. America wanted all POWs on both sides to be repatriated. North Korea wanted Americans who wished to stay in North Korea to be allowed to do so. There were only two or three such Americans. Probably ten thousand Americans were killed, and countless others were wounded over that issue. What silly games we play.

At one point in Korea, we had a trainee tour guide named Queenie. One day I asked Queenie if I were to return to Seoul after the China trip, would she be my tour guide to other parts of Korea where I had been during the war? She said that she would.

China Leg

As the others flew to Hawaii, we flew to Hong Kong to begin our tour of China. Throughout the China leg of the trip, our five, along with six others, had our own private tour guide. She was a very cute little Chinese girl named Monica who wore very short skirts and spoke seven languages. Her Chinese name is Jin Hong Yo. Her first name means "Golden Sunshine," and her family name means "World," so she was Golden Sunshine of the World, and she lived up to her name. Monica was in charge of logistics. She spared no effort in keeping us comfortable, informed, and entertained. Before the trip was over, she seemed like one of us; and I think she enjoyed the tour as much as we did.

Monica

Monica showed us amazing places, and she had a thorough knowledge of all of them. She was easy to talk with, and we all left at the end of the tour with a much greater knowledge of China. I came to love China and the Chinese people.

The entire trip involved several flights (7 in all) that took us to various cities. We saw the terra-cotta soldiers in Xi'an and took a boat ride with the Chinese junks stopping off at a huge upscale Chinese restaurant for dinner.

We climbed the great wall north of Beijing. We had Peking duck in Peking and saw many shows of people doing almost unbelievable feats.

We visited the Forbidden City (so named because it was forbidden for anyone to enter or leave the city without the emperor's permission), and we visited a jade factory and a cloisonné factory. We visited a cave and had wonderful food wherever we went.

Monica and Dick in a Cave

We visited a major Buddhist temple on Buddha's birthday. Now you think that wasn't a big shindig? Wow! There was every kind of Chinese lantern that you could imagine; they went on seemingly forever. There was Chinese music and incense everywhere.

At one point, I was standing in the midst of a great many people when my foot hit something on the pavement below. It was a tattered, dirty, grimy man who had crawled from no telling how far away to arrive at this temple on this day. I don't think he was able to stand up. He had tortured his body in this way for Buddha's favor. I wish he knew that you only have to believe in Jesus and his death on the cross to be eternally saved by the only true God.

They were about to re-roof the temple and had a lot of terra-cotta roof tiles laid out, and people were invited to sign them. I signed one thinking, "Ha! They are going to put this on their roof, and a little beam of Christian light will shine down amongst the evil spirits, irritating and aggravating them for maybe 50 years to come."

Now Carl had a store in Greensboro, North Carolina, where he sold art objects and other things. He wanted to buy some of the jade, but he had to buy it at a good price. I helped him, and we negotiated a deal where he would pay one-third of the retail price. They opened the store for us again that night, and Monica took us back. They had all of their jade displayed for us, and we had a ball making (not ball-making) selections. Carl would buy whatever Monica and I liked. The deal was closed, and they were to ship everything to Carl.

We did the same thing at the cloisonné factory, but this time, after all the selections were made and it came time to pay, they wanted Carl to pay for shipping. He really wanted the goods, but he didn't want to pay the cost of shipping. We talked together until I finally told him, "If you are not willing to walk away, you are not negotiating; you are begging." He was afraid to possibly lose the deal if he walked away, but he finally agreed, and we went to the bus, and the driver started to back out. That's when the sales crew ran out to us and agreed to pay for shipping.

It just happened that after the tour was over, I was in Greensboro to visit my son, Tim, and I also visited Carl at his place. While I was at Carl's house making gumbo, he got a call. It was from the shipping company. His art objects had arrived. There were two large crates from China at the port in Wilmington. Carl was advised that he should pick up the crates rather than have them sent by truck, so off we went in his car, pulling a trailer to get them. We had fun opening the crates and unwrapping the pieces. Not a single piece was broken or even chipped.

Carl told me later that he had done well with the jade but not so well with the cloisonné. He gave me a cloisonné eagle, which I liked and which now adorns the top of my étagère.

We saw fabulous views of beautiful cities and mountains covered with perfectly terraced rice fields. Almost every evening, I called home to tell Rissi about the events of the day. I didn't know what all those long-distance phone calls would cost. It turned out that Rissi could have gone with me for the price of the phone calls.

We took a boat ride on the Li River outside Guilin, where there are many different shaped little mountains on either side of the river covered with vegetation. The mountains all have names according to their shapes. I don't remember them, but they had names like Elephant Mountain, Goat Mountain, etc.

That evening I saw a fisherman on a bamboo raft. The raft was about 4 feet wide and about 14 feet long and made up of pieces of bamboo about 6 inches in diameter. The front was turned up at an angle and came to a point about 24 inches over the water. A single-mantel Coleman lantern hung from the bow. The fishermen pushed the boat with a long pole. On the boat was a basket and about a dozen cormorants, each fastened to the raft by a short tether.

One by one, he released the cormorants. They immediately dove in and "flew" underwater. Very quickly, the first one came up and had something about 3 inches long in his beak. His neck was stretched way out, and he looked up as he paddled to the raft. The fisherman helped him aboard, picked him up, and turned him upside down over the basket. A fish about 18 inches long fell out. It was as long as the cormorant's neck! Soon the basket was full of fish. Later I got on the raft and checked out the cormorants. They are much more beautiful than the ones we have here. Ours is black; theirs are dark but have overtones of red, blue, and green.

My Trip Back to Korea

So, the China trip was over, and I flew back to Seoul and met up with Queenie. The first place we went was to Koji-do.

I saw next-to-nothing of the camp where we stayed, but there is a small museum on the site. In the museum are some pictures of the compounds, and there are some of the prisoner's weapons and some stories of the captured and disgraced General Dodd. The information must have been pulled together long after the war because some of it was incorrect. There were some errors in the written documentation, especially the one calling compound 76 by some other number. They are correct now. I took the liberty of correcting them.

The weapons that the North Koreans used were on display there, including the wooden rifles with bayonets fixed and the homemade hand grenades and Molotov cocktails that I was familiar with.

One day we were downtown visiting a little shop. Queenie picked up a figure of a little seated fat man figure and called it a Buddha. I told her it was not a figure of Buddha, but she insisted, and I insisted. She called the sales girl over, and she also said it was an image of Buddha. I asked for the manager to be called, and he said I was correct, that it was one of Buddha's prophets. I had learned that some of the little "Buddha" images you see have curly hair, and some do not. Only the curly-haired ones are Buddha.

Queenie told me that in a place like this (Koje-do is an island in a pristine sea), there are raw fish restaurants. I said, "But, Queenie, we had sushi last night." But she said, "No, this is different; it is all kinds of raw fish." She said she had never been to one and she didn't know how to find one. So, I said, "Well, I do." I looked around and spotted a man in a business suit, walked up, and got his attention so Queenie could talk to him. In a minute, he stepped out into the street, flagged down a taxi, opened the door for us, and told the taxi driver where to go. In no time, we were at a raw fish restaurant by the sea.

The place was like an aquarium. There were many tanks with many different kinds of fish. There were some people there eating live fish. The fish are taken from the tank, fileted but leaving the intestines intact, sliced to bite-size, and laid on a bed of noodles with noodles also covering the exposed bones and intestines. They are eaten while the fish gasps for breath. I didn't order that one.

I didn't want to select a fish and not be able to eat it, so I asked about a seafood sampler. I thought surely I would be able to eat some of it. They told me they could do that, but it would be very expensive. For each fish, on the sampler, another fish had to be killed. They did fix me up with a fish and some small sides. I was able to eat some of everything. Queenie polished off the rest.

Queenie was having trouble with Christianity. The sticking point for her was she couldn't believe a perfect God would allow so much evil in the world. I tried to help her and got Rissi to send her an Amplified New Testament, which I hoped would help her.

Korean War Monument

I read that the Korean War Monument in DC was to be formally dedicated and that there would be a computer there that could be used to look up people who had fought in the Korean War. This was in the very early days of computers. I went to the dedication primarily to look up my old buddy Cpl. Hammond. Several other 10th Company Rangers were at the dedication too.

The computer was not in service yet. President Clinton was there. He made presentations to the granddaughter of a highly decorated South Korean Soldier and to the grandson of Cpl. Hammond by a son he had never seen. How about that?!

We also viewed the monument itself. The statues of the soldiers on patrol were very realistic to us. The wall was interesting, but it was very crowded, and we were rushed through too fast. So I told the guys that I was going to get up at 3:00 the next morning and catch a cab back to the monument. I wanted to view it more slowly, and I wanted to see it at sunrise.

All the guys said, "No way!" They were not going to get up that early. Then one of them said, "I'll go with you." Then another said he would go, and finally, all agreed. It was all we could have asked for. The sun coming up behind the American flag was beautiful. We had the whole thing to ourselves, it was quiet, and we saw the wall and the patrol at our own pace.

Art and the Maine Reunion

Art Balominos hosted a reunion in Mystic, Connecticut. Art used to call me out of the blue and say, "This is Ott." "Who?" I'd ask. "Ott from Yak, Maine." They talk funny in Maine.

Those of you who have been to Mystic Seaport know that it is a charming old seaport that displays much of the old whaling days and caters to tourists. Rissi and I enjoyed it immensely.

Ott, excuse me, Art took us to a Smith & Wesson factory in Houlton. That was very interesting to me. At the entrance, there was a huge standing mount of a grizzly bear. Wow! That guy was big. We watched while a block of steel was made into the housing of a revolver. Amazing! We also saw automatics and pistol parts being produced. A few of their guns are made with much more care and precision and to closer tolerances than normal, and they sell for quite a bit more. Some have intricate designs which are actually carved by hand. We watched while some received intricate designs cut in freehand.

Art took us out on a boat to see whales. It was great to watch the huge creatures make the big rolls with their flukes high out of the water.

After the reunion, Art invited Emmitt Fike and his wife, Luby, and Rissi and me to his home in York. His home was actually outside of York and right on a bay just off the ocean. His home had been a two-story cottage. He added another story by hand-digging a basement. He hauled the dirt out in a bucket. Imagine! Not satisfied with that, he added a fourth story on top and then a lookout above that. The access to the lookout was a small circular stairway around a pole. He created this five-story home, and he did every bit of it by himself. When the Ranger tackles a job, you better get out of the way.

Art took us down to the docks, where the lobster fishermen come in with their catch. We bought lobsters enough for everybody, took them to Art's house, and he showed us how to boil them. He also showed us how to eat every last bit of a lobster, even the tiny parts of the legs. He told us that their family was very poor when he was growing up, and about all they had to eat was lobsters. His dad did not allow them to waste any part of them. Most of the lobsters we had were normal size, but we had one that was huge. Big six-foot-four Emmitt got the monster and was barely able to eat it all.

Trip to San Diego Reunion

Ranger Emmitt Fike was a good friend in training and in Korea, and he became an even closer friend after the war. He served for a while in a counter-fire platoon with a bunch of straight legs. That is very hazardous work because it exposes you to the enemy. Fike had loaned a pair of his boots to a soldier who had lost his. Later that soldier was blown up badly. His dog tags were missing, so they couldn't identify him until they saw the name Emmitt Fike written inside his boots. So Fike's parents were notified that he had been killed in action.

For a reunion in San Diego, Emmet and I decided to do something different. We talked about taking a train to San Diego; he would take one from Warrior, Alabama, and I were to join him at Lake Charles. We vetoed that because it was difficult to schedule and because it cost more than flying. So, we decided we would both fly with our wives to El Paso, where we reserved a car to drive through Old Mexico to San Diego.

When we got to El Paso, the car rental place there refused to rent the car to us when they found we would be driving through Mexico. So, we rented an old Mercury across the border and headed out on our trip. It was the middle of summer, and about 10 miles out of town, the air conditioner went out. We rolled down our windows and continued on.

Now there are two roads we could have chosen for the trip through Mexico. There was one with good roads that stayed fairly close to the border. The other was south across the Chihuahua Desert to Chihuahua and west across the

Rockies. People told us we couldn't go that way. The map did show a small break in the road at the Continental Divide, but I told everyone, "That's not a very big gap; we will find a way through it." So off we went to the town of Chihuahua (where Pancho Villa had lived) and headed west toward the Rockies.

We spent that night in the little town of Basaseachi. There was a waterfall there that I wanted to see, but the river that fed it was dry. The hotel was run by a little lady named Virginia, pronounced Bitahenia. The hotel only had three rooms, but we only needed two. The rooms had dirt floors and lace doilies on the chair arms. Bitahenia told us that there would be no hot water. I said we have to have hot water, so she said, "Okay, I'll start up the water heater." Rissi noted the screen door type latch on the door, and Bitahenia said we should not worry, that she was going to walk back and forth in front of our rooms all night with a shotgun.

Bitahenia told us that we would not be able to go across the mountains in our Mercury. The roads are too rough for a car, the rivers we had to cross would flood if it rained and anyway, there are no gas stations. "Are there farms along the roads?" "Yes." "Do the farmers have tractors?" "Yes." "Well then," I said, "We will buy gas from the farmers."

The rest of the trip was interesting. There were lots of trees with very red bark. That's not part of the story, but it was interesting. The road became increasingly rough. One time the road was so steep and rough that I thought we might not make it, but there was a guy there with a mule, so I thought he could pull us through if we needed it. Anyway, we made it.

Another time a river ran across the road. We waded out to see, and it was a little too deep for the car to cross, so we carried rocks into the river to make it shallower. The road became narrow, with the mountainside straight up and the other side straight down. A long, long, long way down. The roadway was under construction, and big trucks had gone through making ruts so deep that if we drove in the ruts, we would have been high centered. I had to drive on the cliffside of the ruts with my wheels between the rut and the cliff. My wheels were just inches from the drop-off. I drove like that for many hours.

I had to climb up a steep hill to a farmhouse, where I asked a lady if she would sell me some gas. She asked her husband, who said when he finished his supper, he would sell me some.

Another time a young girl came out with a big glass jar of gas. It looked like about a 25-gallon jug, which she carried on her shoulder. When she got to the car, she put the jug down, spread a cloth on the car's hood, and then put the jar on the cloth, took out a length of tubing, stuck the tubing in the gas, sucked on it, and siphoned it into the car.

I had done all of the driving up to that time, and Rissi told Emmitt he should spell me off. It was at night, and in less than an hour, Emmitt went to sleep, and we drove off the road. Luckily, we were at a place where the land was flat, so we were all okay. I drove all the way after that.

There was a time when we drove over 150 miles down a dry river bed at night. I had everyone carefully watch both banks. I didn't want to miss the place where the river bed roadway turned up out of the riverbed. When we finally found it, we were in a very small village. We had crossed the Great Divide, and the road was paved, but we still had a long way to go.

After a time, we were all exhausted. I pulled off, and we all fell back in our seats and would have been asleep in two minutes, but as luck would have it in two minutes, the dome light turned on with a loud click. Two minutes more, it clicked off. That repeated until we took the lamp out of the light. No more light, but the loud clicks continued.

None of us could sleep. Emmitt had all he could take. He got out, walked back down the road behind us, and lay down on his back in the middle of the road.

Later his wife, Luby, asked in a slow, extended, weak, dragged-out voice, "Dick?" "Yes," I answered in the same voice. "Do you think if a truck came over that hill back there, it would run over Emmitt?" Same voice. "Yes," same voice. "I think so, too," same voice. Emmitt was on his own. We hoped he would be alright.

So, we finally got to Hermosillo and went to a hotel very early in the morning. Man, were we wanting a bed; but we wanted coffee and breakfast even more. They had just opened the little coffee shop, and the coffee wasn't made yet. "One minute, sir" was all we could get from the waiter. A couple more "un momento," and Emmitt was through waiting. He asked me how to say "right now" in Spanish. I told him "Ahorita," and he went to find the little waiter.

Now Emmitt was 6' 4" and way over 200 pounds of muscle, and was a very imposing figure with his full dark beard. He picked up the waiter by his shoulders and said, "CAFÉ - AHORITA!" We got our coffee "right now."

The next day we were on the way to San Diego when we came to a mountain where the road made a very long curve to the right. We could see the whole curve, probably a couple of miles wide. The road was one of those narrow straight up and straight down ones. This one was way, way, way down. We could see trucks, busses, semis, and four-wheelers that had not made it. No one could have survived. That was a little scary.

So, we made it to San Diego, and everyone enjoyed the tales of our trip getting there. We took the easy way back home.

New Red Jeep Cherokee

Carl came to see me in Lake Charles in his new red Jeep Cherokee. He planned to visit his good friend Ranger Bobby Turner in Hemphill, Texas. I went along when he left for Hemphill.

Bobby was something. He lived with a Mexican woman named Rosa, who was a wonderful cook. I watched Rosa as she cooked, and I learned a lot about Mexican cooking. He said that he had known Rosa all of his life and that she had stayed with him through three marriages.

So, then we decided we would all make a trip through Old Mexico. We were just three guys out looking for a good time.

Carl wanted me to drive as he always did when we were together and I am always glad to do it because I love to drive. This time though right from the start, I felt uneasy about the trip. I drove very carefully because of that. I continued to feel that way every day.

Back when I was making trips to Mexico to fish bass or to shoot doves, my kids all asked me to take them on a trip. Mel wanted to fish Lake Guerrero. Steve wanted to shoot doves, and Tim just wanted to see the sights. So, I had to go on three trips instead of just one. I had taken Tim to see Horsetail Falls.

Now I was taking Carl and Bobby to see the falls. We had to ride horses or mules to get to the place where the falls are viewed, which was on the other side of the canyon from the falls.

We had fun riding up the mountain and considered ourselves real cowboys by the time we got where we were going. When you arrive, you walk onto a very wide-open grassy area. You view the falls from across a gorge and you can only see the top of the falls when you arrive. The land slopes downward very slightly, and you walk forward to see more of the falls. The land gets steeper and steeper as you try to see the bottom of the falls.

Finally, without realizing it, you are on a very steep slope. At that time, one little slip and you start rolling, and there is nothing to stop you. Many people have died viewing Horsetail Falls. Now there is a low railing that warns you when it is time to stop.

We all had a great time, but I continued feeling somewhat apprehensive about driving. We ate a lot of Mexican food, including cabrito, which is a young goat that's stretched open and roasted on an open fire. We also learned to eat chilepeqin, a really hot little pepper that grows wild and is harvested by women who sell them in little bags. You take a bite of food and pop in a Chilepeqin, bite down on it, and it gives you an oomph. That's the way it was described to me by a Mexican restaurant owner who first gave it to me while on a white wing dove shoot years before. It is a very apt description.

On Sundays, some villages in Mexico that are on a main thoroughfare conduct sales along the highway that runs through their village. Shelters are set up, and wares are displayed on tables. It is a festive time. They fly flags, musicians play Mexican Songs, and sometimes there is dancing. There is food to be had and a great time is had by all. Carl, Bobby, and I were going through one of these villages and wanted to stop and see the sights. Carl said, "There is a place where you can pull in and park." But something told me I should drive all the way to the end of the cars and park there. From there, we could walk the whole length of the cars and shelters and back. After we had walked the length of the displays and were almost back to our Jeep, we heard a loud crash, followed by another crash and another. They were getting closer and closer to us. Finally, the last crash and the last car ended just short of Carl's Jeep.

There were people lying in the street bleeding, others unconscious in their cars. Kids and women were crying. Men immediately started caring for the injured. We all looked at each other and asked, "How can we help them?" The answer was, we couldn't. We would only be in the way if we tried. So, we got into the Jeep and went on.

The funny thing was that after that, all my apprehensiveness was gone. I sped on down the road like I was invincible.

Was God telling me there was danger ahead? Was it Him that prompted me to pass up Carl's suggested parking place (where we would have been hit)? I think so.

Emmitt's Limousine Bull

I often visited Emmitt and Luby at their home in Warrior, Alabama. He lived on a big wooded site with a pond and big pasture in front. There is a deer blind in the back overlooking a little cleared area where he planted something to attract the deer.

His house is rustic with a big stone fireplace in his den that he built himself. His den features a big oversize Browning Automatic Rifle hanging from the big beam in the overhead framework. Countless rifles, deer head mounts and the like are everywhere. The den is totally masculine, and it is where he and Luby spend most of their time. Rissi and I once got a really nice Waterford crystal biscuit barrel for Luby, thinking there should be at least one thing of her's in that room.

I don't know what she did with it, but we never saw it in the house.

Emmitt had some strange-looking equipment in his workroom and told me he had bought Ranger Ralph Abbot's knife sharpening equipment.

Emmitt made the knife sharpeners and set up shop at gun shows. It would take him less than a minute to get a knife sharp enough to shave. He used a Baldor 3600 RPM grinder with two wheels, one for sharpening and one to buff the wire edge off. The knives are sharpened without using a tool guide and with the blade faced down. He would sharpen knives for a dollar per knife, and he sharpened hundreds of knives a day. He also bought and sold guns, so a three-day gun show was very profitable for him. I bought four grinders and got him to make me four sets of wheels for my kids and me.

Ralph had told me about his knife sharpening. He made up this big story about how he made the sharpeners and how he sharpened all kinds of knives and even scalpels for the local doctors.

One day Emmitt took me on the back of his three-wheeler to feed his 30 or so cows and one very large red Limousine bull. He bought old bread to feed them and had about 20 loves with him.

After he fed the cows, we got back on the three-wheeler, and he began chasing the cows. Then he chased the bull for a while and turned around and let the bull chase him.

That went on for a while, with me bouncing around on the seat and trying to hang on for dear life. I could almost feel the bull's hot breath on the back of the neck.

Fort Benning with Mel

Melanie lives with me now and we travel together when we can. One time when she and I were returning from a visit with Tim in North Carolina, we stopped off in Columbus, Georgia. I wanted to show her Fort Benning, where I became an Airborne Ranger.

Fort Benning had changed a lot. When I trained there, you could see the 250-foot jump towers from anywhere on the base. Now there are so many trees and buildings that the towers are difficult even to find.

We went to Ranger field, which owes its very existence to my friend Ranger Emmitt Fike. After the war, Emmitt and Luby located and went to the gravesite of every Ranger killed in Korea that he could find. He rounded up nearby Rangers, and they performed a little Ranger ceremony with the deceased's family and installed a Bronze Ranger Tab to the headstone. He did all this at his own expense.

Ranger Tab

Emmitt also met with the top Brass at Fort Benning and got them to approve a location for a Ranger Monument. He picked out a rough site. The Brass said, "You don't want that location. It's rough and uneven, and it has a deep drain running through it." Emmitt said, "That's why I like it. It's Ranger territory."

The entrance is paved with bricks, each bearing the name of a Ranger. We saw Emmitt's brick front-and-center and abutting the side walk with mine close to his. He had seen to it that mine was near his. We found many bricks bearing the name of our 10th Company Rangers a good many that Mel had met and known.

There are several monuments at Ranger Field. We read the monument of Korea's Rangers with the names of 10th company Rangers who were lost during the war. We read the

Ranger Creed engraved on the huge stone. We saw where Emmitt was inducted into the Ranger Hall of Fame.

Emmitt Being Inducted into the Ranger Hall of Fame

We saw where a few of us gathered numerous times to spread the ashes of one of our own in the "Missing Man" ceremony. We would line up in formation, and the leader would call role. Each Ranger would answer "here." Then the deceased Ranger's name was called. There was no answer. His name was called again. No answer.

His name was called again, and someone would answer, "He is not here, sir. He has answered the call to a higher service." Then we would all walk forward, fanning out like the strouds of a parachute and spreading the ashes. *Ranger ___ you once flew on wings of silk. Now you soar on the wings of eagles.*

We went to the big museum on the base, where I showed Mel a picture of Medal of Honor winner Cpl. Hammond. She said "He looks baby-faced". "He was", I told her "but what a soldier."

I showed her the picture of Ranger Hall of Fame recipient Emmitt Fike. I started reading to her the words about him and what he had done. After I had read for a while I realized they were the words I had written in recommending him into the Ranger Hall of Fame.

I recommended him for all the work he had done in installing the Bronze Ranger Tab emblem on fallen Ranger's headstones. Luby went with him on all those many trips to meet with the families. I also recommended him because he got the Ranger Monument established and built at Fort Benning.

There are sections there where every American war is depicted. There they show the arms used, the uniforms, and stories of each war. The Korean War is the exception. There is nothing about the Korean War in this museum or in any other museum in this country. That is because it was a war that the government did not want to admit to having. It was called a police action even though there were almost as many

Americans killed in three years in the Korean War as in the Viet Nam War, which lasted six years.

Anyway, needless to say, the visit brought back a lot of memories.

Chapter 6: Business

Start of Construction Business

Rissi had always thought that if she married me, she would be living in a houseboat on the river somewhere. Instead, I was in living in Aransas Pass, Texas, and working as a roughneck on an oil drilling rig.

When we got married, I moved from Aransas Pass to Corpus Christi so she could work as a nurse in a hospital there and I kept working on the rig. After some time, I got tired of paying $90 a month for apartment rent and looked into buying a house. At that time, I could buy a small home on a VA loan with payments only slightly higher than the rent we were paying.

When I was about to pull the trigger on the house, I told her, "If we do this, we will be putting down roots and will probably be living in Corpus Christi permanently. Is that okay with you?" She said she would like to move back to Lake Charles someday, and I said, "I would too, so what are we doing here?" We both quit our jobs, packed up, and drove to Lake Charles.

Rissi worked as a nurse while I built a house for us. Except for a two-week period when my brother Phil helped me while he was on furlough from the Army, I built the house by myself. I poured and finished the concrete, drove every nail, laid every brick, and put on every shingle. I put in the plumbing and did the wiring. However, I did hire a company for the linoleum flooring and for the Formica cabinet tops.

I didn't know how to do all that when I started. When I would get to something I didn't know how to do, I would drive around town, find a house under construction where that work was being done, watch the workman, and learn how they did it. It was a slow process. It took me nine months to build that house, but I learned a lot.

Then I built another house, this one for sale, and hired one man to help me. For the next house, I hired two people to help me. After that, I hired others to do all the work.

Soon I was asked to do small commercial jobs, then larger ones. Then I expanded into industrial work at the local petrochemical plants. Eventually, I decided that home building and commercial building were not compatible, and I concentrated on commercial and industrial work.

For many years, I ran a successful construction business and built many of the commercial and industrial buildings as well as churches and schools in Lake Charles; but I give most of the credit for our success to the wonderful people who worked with me.

People like Ray Bourque, who did most of our estimates for a lot of years. Ray and I also fished together. We made numerous trips to Lake Guerrero, Mexico, with our key people. On one trip, we were crossing the lake, far from shore, when we spotted a large land snake. We knew it was a land snake by the way it was swimming. It was a really big snake, about half again as long as the boat we were in, and it swam with almost two feet of its head and neck straight up out of the water. I grabbed a landing net and told the guide to go up to its head, so I could catch it. Ray came unglued. He fought for the net and control of the boat. He kept me from landing it.

One day Ray and I were fishing for bass on the river when we came across a huge wasp nest full of those big red wasps. Right next to the nest was a bright shiny topwater plug. I was in the bow of the boat, so I got Ray to paddle me up really slowly to the nest. He did, and very carefully, I unhooked the plug, and we backed out. A little later, we met up with a friend of his who told him he had just lost a brand-new plug when it caught next to a wasp nest. So, what does Ray do? He says we found it, and he gave it to him. What? It might have been the friend's because he bought it, but it was mine because I risked my life for it. Oh, well. I let it go.

And then there was Betty Oakley, who was my bookkeeper for many years. But she was more than a bookkeeper. She would advise me and tell me when I was about to make a mistake. We furnished Mercury cars for our coordinators and over the years bought a lot of them. One day I tried out a new one at the dealership and told them to send me a bill if I didn't bring it back. A year later, Betty discovered we had never paid for it and told the dealer. Thanks a lot, Betty. Well, I guess if she hadn't, I would have.

There were men like Lee Bruney, Red Leger, and Nickie Priola who ran our crews and Snooky Busby, whose workmanship no one has been able to copy (and many have tried) and Terry Leger who could do more demolition by hand and drive more stakes in the ground than anyone, and hundreds of others. I shouldn't have started naming; there is no place to stop.

And then there was Lee's son Lee Allen Bruney who later led the work on the American Embassy Building in Moscow, Russia, and after that became a successful businessman and a foremost leader in our community. I am proud to have had a part in heading him in the right direction.

Blimp Factory

We built a lot of buildings for the drilling company Magcobar, mostly drilling-mud warehouses, and office buildings. One job was to be in Houma, Louisiana. Magcobar was using an old blimp factory building for a mud warehouse, and they wanted me to build a new warehouse

15 feet from the old blimp factory and then demolish the old building.

The blimp factory was huge, something like 90 feet wide and 200 feet long, clear-span and high. It was built with enormous wood timbers bolted together to form trusses. Some of the timbers were like 6 x 14 x 30 feet. I contracted with a demolition guy named Al to take the factory down. Al and I made a deal. I was going to get the timbers, which he would haul to a property I owned on Moss Lake, and Al was to get the miles and miles of big copper wire. Each of us was to come away with salvage worth a LOT of money. Al first removed the wiring, piled it up, and burned the insulation off.

I was there the day Al started to take down the structure. He began by removing the purlins that ran between the first two bays. When there was only one left, he put a winch line on the first bay to pull it down. I said, "You are not going to cut the last one?" He said, "No, I'm just going to pull the first bay and break the purlin." I thought he should have at least cut it part way but he didn't. Well, when he did, the whole great big building came down like a row of dominos. The timbers were so old and dry that they were reduced to splinters. The morning Al went to pick up his wire, he found that someone had hauled it away the night before. Neither of us got our salvage.

Luckily, the building collapsed inward, and only a couple of sheets were dented on the new building. Thank you, God.

Rice Scoby

Rice worked for Magcobar. I worked for Rice. He would call me and tell me to meet him at his home the next morning. We would have coffee, then drive to the airport and arrive just in time to board. Rice hated to wait in lines. We repeated that a good many times.

Well, you know what finally happened. We missed a flight. Rice was beside himself. He had set up a meeting with his boss, his boss's boss, and his boss's boss's boss. They were all flying to Houma to meet him. He said, "Dick, I don't know what to do."

Dick to the rescue. I said, "Well, I do. We are going to McFillen Airfield and find somebody to fly us there in a private plane." We did and arrived sooner than the commercial flight did because it had a stop in Lafayette.

A typical trip was like this: Rice and I would board a plane, and I would ask, "Where are we going, and what are going to build?" Rice would order a drink and draw a plan of the building on a napkin.

One time he asked me to have a crew of men at Harvey in the morning, and we flew down to meet them. The building was to be a drilling-mud warehouse near the river. Because the ground was unstable, pilings had already been driven. When a tug boat went by in the river, we could feel the ground shake from 300 feet away.

We also built a mud warehouse in Grand Isle with an office on one end, and he asked me to design and build a boathouse there on the water to house two 45-foot boats. I shouldn't be designing such a thing. I am not an architect or an engineer, but I did it. We had just completed the work when a hurricane blew in. Grand Isle was hit hard. The building was built on sand, and sand had washed out from under the corner of the building, leaving a 20-foot by 20-foot section of the building hanging out with nothing supporting it. We packed damp sand under that section to keep it from sagging, and we made extensive repairs to the building. The boathouse? It was fine—no damage whatsoever.

W. T. Burton

W. T. Burton was a Lake Charles business tycoon and philanthropist. There is a larger-than-life bronze statue of him in front of the Lake Charles Burton Colosseum that shows him with the Wall Street Journal folded under his arm, the way we often saw him. Mr. Burton had a lot of land and all that goes with it; oil and gas production, cattle, you name it. For years he had the rights to all shell dredging in South-

West Louisiana. There were shell banks up and down the Calcasieu River, and I don't know where all else. He dredged it until the shell banks were almost a thing of the past, and his shell covered roads all over this part of Louisiana. He also owned the Calcasieu Marine National Bank, which is the largest bank between Houston and New Orleans. I got to know him when I did my banking there.

W. T. Burton

Later I moved my accounts to Lakeside Bank and served there for many years as a bank director and as chairman of the audit committee. One day I saw Mr. Burton and congratulated him on a milestone his bank had achieved in total footings. He punched me repeatedly in the chest with his finger while he said, "Yes, but we didn't make it on the backs of the ones that left us to go to Lakeside."

Wow! Mr. Burton didn't like losing a customer.

New High School

I once had a contract to build a new school building, complete with an up-to-date cafeteria and a gym. After the wood gym floor was installed, we had a hard rain. The roof leaked, and the gym floor was destroyed. There was no doubt the maple floorboards were cupped too badly to just be re-sanded and re-finished. Most of the flooring had to be removed, and some of the sub-flooring and the sleepers had to be replaced. It was going to be a big deal.

It turned out that a small section of roofing around a drain that was connected to the storm sewer had never been finalized. When the roofer got to that place in the work, he left it incomplete because the plumbing drain had not been completed. When the plumber got to that place, he could not complete the work because the roofer was not there. Both the roofer and the plumber blamed the floor installer for not making sure the building was ready for the floor. Everyone blamed everyone else, and no one wanted to take responsibility.

All the parties notified their insurance companies, and all of the companies were taking a "not our fault" stance. It was shaping up to be a big series of court battles. That was going to cause a delay in completing the job, and I would probably be sued by the school board. It was not a pretty picture. I had to do something.

I called for a meeting with all the subcontractors involved and asked them to have with them an insurance company representative who had the authority to make decisions for the company.

We all met together at the school, and I told them what we were going to do. The roofer would pay 70% of the cost to repair. The floor installer, the roofer, and my company would each pay 10%. After a little conversation, all were in agreement except for the plumber, who said they would take 5% but not 10. Now the whole agreement is being held up by 5% of the cost. What to do? So, I told the roofer that my company would not pay another 5%, but if you can talk one of these other two into taking 5% of yours, I will be okay with that. After a few minutes of the three sizing each other up, the plumber said okay, his firm would take 10%. The sub-contractors had done all the talking. The insurance representatives took what we had agreed upon.

We completed the job on time and within budget.

Key People Perks

When I was a building contractor, I took my key people on trips. We mostly went to Mexico for fishing and once to Colorado for snow-skiing. One year I offered them a deal. If we put five million dollars worth of construction in place, we would go to Colorado for two days of skiing. If we reached seven and a half million, we would ski four days. If ten million, we would take our wives. We reached more than ten million, and everyone had a great time.

My job was easy back then because of my people. I had good people who knew the work, and I could trust them to do the work well. They always had my interests at heart.

Dana

Dana worked for me for many years. Dana was amazing. She loved people, and everyone loved Dana. She could talk and relate to anyone, professional people, doctors and lawyers and such, as well as working-class people, housewives, convicts, pimps, prostitutes, and criminals. She knew how to handle outraged customers, and almost without exception, they would go away calm and at least somewhat satisfied with the outcome.

I had an office building (owned by my two brothers and me) where I operated a construction business, a self-service warehouse business, and a mobile home park business. There were numerous other offices in the building.

Dana's office was at the front of the building, where she could greet people as they came in.

Dana at Work

Often there was a crowd of people around her desk of people just to visit with her. She was able to do that and all of her work while directing people who entered the building, overseeing the maintenance of the office building and the self-service warehouses, answering the phone, renting storage spaces, maintaining records, doing the bookkeeping, and a lot of others things. When she had a break from the phone and the people in the office for business, she would go back to the conversations with the people there to visit and pick up the conversation right where it left off. She talked to them while continuing to do her work. Dana always completed her work on time and without error.

"What?!" you say; "Nobody does that." Dana did it, and with ease. And if that weren't enough, she was beautiful to boot - still is.

Many people in Lake Charles knew Dana, and some knew what she was capable of. So, it was not a surprise to me that another firm wanted her. She was offered so much more than I was able to pay that to offer her a raise to keep her would have been futile. She turned down the offer anyway and stayed with me.

Later Dana took over my mobile home park. She ran the park for me better than I had done it myself and improved it greatly.

Dana has made a lot of money for me, and when I sold all my businesses, I made sure that she would have a very comfortable retirement. She could have retired right then but chose to continue to work.

One day Dana and husband David and sons Cody and Logan all accepted Christ as their Savior at their home and were baptized in their backyard pool.

My daughter, Melanie, and I still go out with Dana and David once a month for a nice meal and a time to catch up.

Dana White

Chapter 7: Glasses Ministry

Letters to Tim

This was written years ago to my son Tim when he asked me to write to him. This is what I wrote:

This is to Tim, who asked me to write and tell about my mission trips, how I got started making them, and about some of the trips themselves. As I start writing, many of the experiences are beginning to flood back into my memory, and Tim, I think I may write much more than you wanted. Anyway, I will send these to you in installments. Here is the first.

How did it get started, you ask? I guess it really got started years ago when Rissi encouraged me to go on mission trips. She would like to have gone herself but did not think that she was up to it physically. I had little interest in it at the time, but she mentioned it to me several times over a period of years. Eventually, in 1995 I had the time and the inclination to go when Jim Selfridge asked me to go with his team of ESL teachers (That's "English as a Second Language") to Ketrzyn (pronounced KENshun), Poland.

Poland

I thought that Poland was a long way to go without combining it with a side trip. I wanted to fly to Iceland a week ahead of the rest of the team and do some sightseeing there. I tried to get some of the other team members interested in Iceland but had no luck so I planned to go by myself. Carol Selfridge and Linda Saucier were going on the trip to Poland, and they wanted to make a side trip to Austria and Germany but had decided against it because they had not found a man to accompany them. Well, you know how that turned out, I joined up with them, and we all took the Austria/Germany trip. It was a lot of fun; we saw a lot of castles and took lots of pictures. We visited one of Ludwig's castles on an island in a lake called Prien Lake. How about that. Afterward, a flight to Warsaw took us to meet up with the team.

The idea there was that each member would each teach a class for two weeks. The Polish people are eager to learn English. I taught a group of High School Seniors, and we had a great time together. We taught English every day, and we also made friends with our class members. Toward the end of the first week, we told them how we came to accept Christ as our Savoir and how it had changed our lives. During the second week, in our sessions comprising all classes, the students were given the opportunity to accept Christ as their Savior. Most of my students became Christians that week.

One day before we went, several of the team members met together to make lesson plans. I was having a little trouble catching up with others who had done this before. At one point, Linda Saucier saw my frustration. She stopped me and said, "Dick, don't worry about all this stuff, just go over there and love the people, show them God's love." I have never forgotten that, and I do love the people that we serve as well as those who work with us. I think this is one reason I continue to do this work.

We had some fun times in Poland also. One day after classes, my students showed me around their city, and took me on a bus ride out of the city to see Swieta Lipka, a beautiful cathedral built way back in the woods. It was packed with people; many came in tour busses from other cities and countries in Europe. In the nave were beautiful figures of carved wood, gold, and other materials. A huge ornate pipe organ in the back of the cathedral began to play, and when we turned to look at it, I saw that the pipes were set in a sea of ornamental figures. There were figures of angels, birds, animals, and many others. Then to my surprise, one of the figures began to move, then another and another until almost all of them were dancing and twirling or moving in one way or another.

It was quite amazing. If you are interested, you can google Swieta Lipka for more information.

A simplified version of the story of the cathedral goes something like this: "Back in the 15th century, a man was sentenced to die for his crimes. In jail awaiting his execution, the criminal prayed to Mary, who visited him and gave him a piece of wood from a Lipka tree and a chisel. She told him to make a carving of her and the baby Jesus and to give it to the judge. He made the carving even though he had never done carving before. The judge and others were so taken with the beautiful carving and the story that they set aside the man's conviction and placed the carving in the church in Ketrzyn. The next morning the carving was found to have been miraculously transported into the woods next to a young lipka tree. Each time the carving was returned to the church, it went back to the little lipka tree in the woods. That was taken as proof that this was indeed a miracle, and the lipka tree was elevated to sainthood." The Swieta Lipka cathedral was then built in the remote woods by the little lipka tree. It is considered to be the most beautiful example of baroque construction and holds some of the most beautiful baroque art objects anywhere. Inside the cathedral is a huge carving of a lipka tree.

We also went up close to the Russian border to the "Foxes Lair," campus of buildings built by Adolph Hitler during World War II. The walls of the buildings are about 15 feet thick made of reinforced concrete to withstand bombing. They were also built with powerful explosives within the walls so that if Germany lost the war, it could be demolished before being taken over by the axis powers. Hitler spent most of the war in the fortified camp. It was all built by Polish forced labor, and all the workers were then killed so that no one would know about the place. As always happens, one person escaped, and was able to tell about it. As Hitler and his armies fell, the camp was indeed blown up by retreating soldiers. The fallen walls and roofs are still there for anyone to visit and walk through the structure. If you have read about World War II, you will remember that Klaus von Stauffenberg, one of Hitler's highest officers, tried to kill Hitler by placing his briefcase full of explosives on the floor between his chair and Hitler's chair. Claus von Stauffenberg set the device to explode and immediately left in a fast car to a waiting plane. Hitler got up from his chair before the explosion. He was injured but not killed. Stauffenberg was captured before he could board the plane the plane and was executed.

That Ketrzn trip was a good one. I thoroughly enjoyed it and loved the students, but I was impressed that ESL was not what God had in store for me.

El Tuma, Nicaragua

I did not make another trip until December 2000 when my friend, Dr. Dick Landry, asked me to go on a mission trip with him. He told me someone had given him a box of used glasses and he wanted me to go with him to El Tuma, Nicaragua and try to find someone who could use them.

Dick makes medical mission trips, traveling to various third world countries treating people who would not otherwise have access to medical treatment. Doctors, dentists, and nurses join him on the trips, and they treat many people with a wide range of medical problems. Sometimes carpenters, builders, and others go along to build or repair churches. His trips are also evangelistic. All comers attend church services, and many become Christians.

I realized that I needed to start working on my Spanish again. Years before, I used to take my key people in the construction business on fishing and hunting trips to Mexico and began trying then to learn some basic Spanish. I had gotten pretty good back then and was pretty much able to get my thoughts and questions across. Understanding their response was another thing. Berlitz tapes and Leisure Learning Spanish classes helped some, but I was far from fluent.

I also decided that I would like to take a little side trip before going to El Tuma. I had wanted to go to Costa Rica for a long time, so that seemed a perfect place to go. One day at church, I mentioned the side trip to Lindsay Saucier (Linda's daughter), who was also going on the El Tuma, and she said unequivocally that she was going with me to Costa Rica. That was okay, but I knew that we needed someone else to go, too. I didn't want any raised eyebrows about me going on an out-of-the-country trip with a young girl. I tried to think of another female who might join us on the exploring side trip. A couple of weeks went by with no luck, so I began to pray God would reveal someone to me if He wanted us to make this side trip. The next Sunday, as Rissi and I were walking to our car after church, I mentioned to her that I still didn't know who else I could recruit. She immediately said, "There is Debbie Turner over there. She and her son Jason are going on the Nicaragua trip. I'll bet they would want to go". With that, Rissi caught up with Debbie and told her about the side trip. She promptly said that the timing was perfect; she and Jason were going.

So, I booked our flights to Managua and reserved a car. We planned to drive across the border to Costa Rica the next morning. When we got to the Camino Real Hotel, Debbie checked in some of her medical gear for them to hold for her until next week. We then went to pick up our car and learned that it would take three days to obtain permission to cross the border. We went to bed, not knowing what we were to do.

Very early the next morning, I went to the airport and asked about flights to San Juan and was told there was only one flight that day and it was leaving in 25 minutes. I bought four tickets and rushed back to the hotel, knocked on doors, and told everyone, "Get up and get dressed. We are leaving NOW.

We made the flight.

We rented a four-wheel-drive Bronco. Jason collected brochures of places to go and things to see and do in Costa Rica, and we put him in charge of planning the trip. He filled every minute of the whole trip with exciting activities. We saw several volcanoes, including the famous active volcano Arenal and zipped through jungle canopies and over deep gorges on zip lines. We toured Deer Cave, which took us through spaces so small that I was sure someone would get stuck and through what seemed like miles of waist-high swift running water. At one point, we had to lie down and slither through a tube-like hole that was about 15' long. It was so tight that the only way we could move forward was with our toes. If the floor had not been wet and slimy, I don't think we could have done it. That trip is not for the faint-hearted. We went through butterfly gardens where wonderfully beautiful butterflies are raised. We went on night hikes through the jungle with a guide and flashlights and saw white-nose coatis, kinkajous and jaguarondis, and more. During the day hikes, we saw wild toucans, macaws, howler monkeys, and lots of iguanas. We took boat rides with naturalists down jungle rivers that teemed with crocodiles and all sorts of birds and other animals. We had all the birds and wildlife pointed out to us and their habits explained. We went on a tour that included a speed boat ride through lakes and rivers, a horse-back ride through beautiful country with a rest stop in the woods where we were served various kinds of tropical fruits. Wow! That was a week to remember. While in Costa Rica one evening Lindsay and I were lounging under palm trees on the beach at Tamarindo she told me that she was praying about a difficult decision she had to make. She said she had a great opportunity to go to China as a journeyman for a year. But she had for years felt God's leading her into the medical field and that she should continue her education in that direction. The decision was weighing heavily on her. It seemed to be a really tough choice. As we talked and prayed together, I suggested that the two did not have to be mutually exclusive and that a year off between her studies was not a bad thing. I think she made her decision while on that beach. She did go to China for the year, and as I write this in 2007, she has also received her Nurse Practitioner degree from Colombia University.

A week later, the four explorers arrived back at Managua just minutes before our mission team was due in from the States. I told Debbie to give me her claim ticket for her gear, and I would catch a ride to the hotel, retrieve her stuff and get back in time to meet the team. I ran into the hotel, handed the ticket to the head baggage handler at the front door, and told him in my best Spanish I wanted to pick up the bag. The handler studied the card carefully, looked at me for a few minutes, and then said, "Un memento Signor." With that, he took the ticket to the concierge, who also looked at the card and at me and back again and finally said, "Un memento Signor" and took the ticket to the front desk. I told the receptionist that I only wanted to pick up the bag that was checked there the week before. The receptionist did as the others had done and finally said, "Un memento Signor." By this time, I am wondering just what has happened to the bag. Our next encounter was with the hotel manager. I told him in Spanish and English that I just wanted to pick up the bag. Finally, he said, "Senor, este boleto esta para el desayuno" (this is a breakfast ticket). Debbie had given me a ticket for a free breakfast, which we had skipped the week before. We have all gotten a lot of laughs about that mix-up. After the trip, Debbie gave me a picture of the four of us wearing our zip line gear. The frame bears the words "We Were Here," and on the back is fixed the famous breakfast ticket. It still hangs on my computer wall.

Lindsay Dick Debbie, Jason

Well, I did get the gear, and we all went to El Tuma, Nicaragua, a small village on a little river by the same name, far back in the hill country. Coffee is the main crop, but there is also corn, bananas, and mangos along the way. When we started work we learned that people would walk a long way, some for a couple of days, to get to us. We went with a huge team of doctors, dentists, builders, evangelists, cooks, hair washers, and you name it. The medications, glasses, bibles, and other things were shipped a couple of months earlier on three huge shrink-wrapped pallets. We got word from the locals that they received the shrink-wrapped packages, but when we got there, the bundles were only about half the size that they were when they were shipped. Much of the medicines and all of the bibles and glasses were gone. I spent that trip helping in the clothes closet. It was a good trip overall, though. The medical people saw 3450 patients and filled over 10,000 prescriptions. They had 530 dental patients and pulled 1667 teeth. The hair crew deloused, shampooed, and cut hair for 1000 individuals, 1300 pounds of clothing was distributed, and the best of all 121 people made a first-time profession of faith. Even so, the local people knew that we were to bring glasses and were extremely disappointed not to get them. So much so that Debbie and I decided that we would just have to go back.

El Tuma Return Trip

We recruited Jason, Lindsay, and Jim Selfridge to go with us. We put the return trip together with the help of BMDMI, "Baptist Medical and Dental Missions" who had also supported our first El Tuma trip. They have full-time missionaries Arnaldo and Arourra in Honduras and Nicaragua who help people from the United States put together mission teams. They handle much of the logistics of the trips. Arnaldo and Aroura also have mission houses in Nicaragua and Honduras with space and facilities for teams to be housed as they travel into and out of the countries. BMDMI furnishes teams with ground transportation as well as meals and translators, they locate places to work and eat, etc. Missionary Roberto sees that everything runs smoothly. Each team member pays for his own trip, which includes airfare and his share of the things that BMDMI furnishes.

So, Lindsay and Jason did a bible school with the kids, Jim held services in a tent, and Debbie and I gave out the used glasses and some reading glasses by trial and error. By this time, of course, the lost drugs, Spanish bibles, glasses, and bibles had been recovered. Since Debbie is an RN, we dispensed some of the recovered medicines, and we gave out a lot of the bibles. Debbie also taught a class on birth control to young girls. They knew nothing about such things, and many teen-aged girls there were pregnant. Even our translators and local helpers had never heard of the rhythm method.

I had sent word to the local missionaries that I wanted things to be different than they were on the previous trip. For one thing, I did not want people to have to wait in long lines from long before dawn until the afternoon to be seen. They were to be told ahead of time what time they would have an appointment. Another thing, I didn't want us to stay in the school compound, surrounded by barbed wire fences with guards stationed at every gate. I wanted to live and work in the village with the people. They were able to make all that happen, and we roughed it with the locals. We slept on dirty floors in rooms with no furniture and worked in an open shed. Jim preached in a tent that was set up for that purpose. There was no church in the village, and the local people wanted one and had picked out a location, but the mayor over several villages, including El Tuma, was not inclined to permit it. Roberto, the BMDMI missionary, and I paid the mayor a visit, took him to the proposed site, and pled with him to allow the church. We prayed at the location and left feeling like we had made progress with the mayor. I later learned through Roberto that there is now a thriving church there. The trip was a success. People were saved, they were fitted with glasses, given medicine and bibles, and a church was started. The people loved us, and we loved them.

As we left, one of the local ladies gave Jason a young rooster, which he held all the way back to the mission house in Nicaragua, about a three-hour trip. When he turned it loose, it immediately perched up in a thick tree, and we were not able to catch him again before it was time to return home.

El Tuma Tits

When we began to unload and set up the first day, there were some local people and a missionary with us. In a few minutes, a very attractive young well-endowed girl in a low-cut dress walked by in front of me. I'm sorry. I looked.

The next time she was walking directly toward me. Every time I looked up, she was there. Her dress was not just low cut; I think she pulled it down a little further.

In a little while, one of the local men said something to the missionary who was there who then said to me loudly enough for all to hear, "He said he thinks you like the girl." I said, "What's not to like?" He said again, "No, he said he thinks you like the girl." Again, I said, "What's not to like?" He said it a third time, "No, he said he thinks you like the girl." After that, I said something dumb, and everyone let it drop, but I was embarrassed. It turned out that the guy who squealed on me was the girl's husband who was quite a bit older than her.

Well, he was right; I did like the girl. We were there for several days, and I got to know her well. One day, she told me that her husband was getting fat and that she was cutting back on his food.

The girls there have children at an early age. They know nothing about contraception. One day, Debbie was teaching the girls about the rhythm method. When she talked about stopping sex for days, our girl chimed up, "My husband wouldn't like that." I told her that she cut him back on his food; she could also cut him back on sex; that felt good. Now I'm thinking you got me buddy, you got me good but I got back at you a little bit.

It took me years and years before I realized what had happened. That girl didn't decide to flaunt herself in front of me with her husband watching. He is the one who put her up to do it. He set the trap, and I fell right into it. Man, how I wish I could have realized it when it happened.

If you don't like the title of this story, talk to God about it. I couldn't think of one less crude, and as fitting, so I asked Him to give me one. He hasn't responded.

Our glasses work, though, was not what it needed to be. We had some off-the-shelf reading glasses like you find in drug stores, and we had used glasses, mostly old-style bifocals. You can imagine the difficulty in finding exactly the right glasses on a trial-and-error basis. Afterward, back home, I did some research and found Holland Kendall in Jefferson, Kentucky, who was working on a way for laypeople to fit glasses in third world countries. His method required a huge number of used glasses, a refractor, and a computer to select the closest match on hand for the patient. That method would not work for us. I also took a course in Dallas, TX, to learn to use a simple device for getting a refractor reading of the eye. In theory, the device would work, but in practice, it didn't work for me. I had to find a better way. About that time, Charles Rue started getting interested in what I was doing, and we began to work together to learn all we could. Dr. Mel Gehrig had given me some instructions, and now Charles and I started getting some help from Dr. Rick McGuirt, an eye doctor who belongs to our church. Finally, Rick took us to his office, showed us a handheld refractor, and said, "This is what you guys need." Wow, it was a beauty. To use it, you just get close to the patient, point it at an eye, push a button, and voila, you get the reading. Then you make some calculations and know what glasses are needed. The refractor was made by Nikon. It is handheld, battery-operated, and can be taken anywhere. There was just one catch. The Nikon version we were looking at cost $16,000. Well, that was a bit much, but after a little research, we found one made by Welch-Allyn that cost less, and they had one that had been used as a factory demo that they would sell for 7,000. After a little persuasion, we got it down to 6,000. Charles and I decided that we would be a team and buy the refractor together. Our church, Trinity Baptist in Lake Charles, LA, began to foot the cost of glasses for us.

We were also able to locate a place to buy separate lenses and frames, so now we had the capability to get a reading on a person's eyes and select the correct lens that was needed for each eye. We then set the lenses in the frames, and by purchasing some cylindrical as well as spherical lenses, we could also adjust for astigmatism. After a little practice, we thought we were ready to go.

Jim Selfridge and I attended a BMDMI meeting in Hattiesburg, Mississippi, to learn how we could better work together on future mission trips. We found Arnaldo and Aurora there. You will remember that they are the permanent missionaries in Nicaragua. Some of the first El Tuma trip members were also there. I asked Aurora how the rooster was doing, and she laughed and said that everyone she knew was asking about the now-famous rooster. She shared the surprise that the "rooster" was now laying eggs.

Kenya

Dick Landry had a Medical and Dental trip planned for Kenya the following summer, so Charles and I signed on. Talk about a long flight. We drove to Houston and caught a direct flight to Amsterdam, where we had a three-and-a-half-hour layover. We still had a twelve-hour flight to Nairobi ahead of us. Amsterdam was a good break for us. They have wonderful Dutch chocolates there, so I bought a big box and passed them out to our team. When none of us could eat any more,

I shared with other travelers until they were gone. The last leg of our flight seemed to take forever. Let me tell you; the Sahara Desert is HUGE. For hours on end, we could see nothing but desert. Finally, we made it to Nairobi, loaded up, and headed for our hotel. It was getting late, and we really looked forward to a good night's sleep.

The first thing I noticed on the way to the hotel was that our vehicles didn't stop for red lights. Nobody stopped for red lights. We were told that it is to prevent being overtaken at a red light and robbed. During the day, the problem lessened, and stoplights were obeyed. The plan there was for us to stay in our hotel at night and make day trips out to small villages around Mount Kenya. We spent the first day relaxing from the trip, then the next day boarded several vehicles and headed for Kanunga and the little Baptist Church there. Daniel Theuri is a kind of senior pastor who oversees and assists many small village churches. Daniel is a well-educated Kenyan who speaks with a British accent. He and his wife proved to be delightful people and were a great help to us.

Charles and I now had our refractor, and it was proving to be a big help. We went to a different little church almost every day, so our workdays were short. We had to spend time every day getting to the church and setting up, and we also had to leave in time to get back to the hotel before dark. No one wanted us on the road after dark. The doctors and dentists got their choice of facilities, and our glasses crew took what was left. The first day we quickly sized up the space available and knew we would not be inside, so we moved into a little open shack with a dirt floor and went to work. Later some ladies with water jugs on their heads let us know we were in their kitchen where they wanted to build a fire in the middle of the floor to boil water for drinking. They quickly agreed to build their fire outside and let us continue to work. Later it got so hot in the shack that we moved outside in the shade of the shack, shade that was slowly disappearing. Many people were sick and said they had malaria. Some probably had HIV or AIDS, but everyone said it was malaria. One little girl finally got to the head of the line, and when we tried to talk to her, she bent over and spit up. We then realized she had a high fever. Of course, we got her to the medical people, who told us later that she had pneumonia and had been treated for it. Debbie, who is an RN was on this trip with us, working with the medical people.

Debbie with a New Friend in Kenya

At another location, we had a good place to work. The people even covered the little building with black sheet vinyl so we could use it for a refracting room. We also had a couple of young men to help us. That morning Charles told me that he was doing nothing but refracting, and he knew nothing about how I was working and wanted us to be able to trade jobs. We decided that he would refract about 15 or 20 people and then come watch me, and I would explain how I made the selection of glasses. Right away, he took exception to what I was doing and became very angry and refused to work with me anymore. He told me angrily that we were going to have a talk later and that he was going to do most of the talking. He was so convinced that I was wrong that he would not listen to an explanation. Steve James, our pastor from back home, was with us did the refracting for the rest of the day.

The next day I got with Charles, who was ready to talk, but I said it had to wait until we could be by ourselves and could sit down together. I thought I could dodge a punch better from a sitting position. When we found the time and the spot, I said that I wanted us to establish some ground rules before he began. He was a little taken aback and asked what rules I had in mind. I told him that first, we should take turns talking and listening, he reluctantly agreed; second, I suggested, no one raises his voice, he okayed that as well, and third, we should agree on a desired outcome of the discussion. When he asked me for my desired outcome, I told him it would be a restored peace between us. He had to agree, and all the fight was gone. He then let me know that he was tired of refracting and wanted to give out glasses. So again he refracted some of the people and watched me as I showed and told him how I did it. Peace was restored. Then we started swapping the refracting job. He still thought I was wrong about one thing.

We had about three days off in Kenya, and they were all interesting. One day we went to the equator, where there were many little shops with local artifacts. Back home, David White had asked me to get him a warthog tail, so I shopped for one. Everyone said that it is illegal to buy or sell such a thing, but one lady whispered to me that she could get one, have it tanned and ship it to me. The cost would be $40 American, and I would have it within a couple of months. I didn't bite.

There was a marker right at the equator where a geologist on duty there showed us an amazing thing. He had a jug of water, a pail, and a funnel with a small spout. We all took 20 steps north of the equator, and he poured the water into the funnel and put a match in the water. The match turned counter clockwise as the water drained into the pail. Next, we went 20 steps south of the equator and repeated the process. You guessed it; the match turned clockwise. Right at the equator, the match did not turn at all. I watched him carefully to try to discover a trick but couldn't detect one. I even stirred the water in the opposite direction, but it would stop and reverse before the funnel was empty. It's called the Coriolis Effect. Since that time I have seen the same thing at the equator in Equidor.

That evening we went to a game reserve and stayed in a place called "The Ark." It was built to resemble Noah's Ark but was actually a hotel. One end of the first floor was all glass and looked out onto a watering hole. The second and third floors had large porches that also overlooked the watering hole. There was also a ground floor "quiet room" with small openings without glass. We could watch from any of those vantage points as animals came to drink. That evening we saw elephants, rhinos, cape buffalo, warthogs, hyenas, and a lot more. Naturalist Joseph Mutongu was available to tell us about the animals and to answer our questions. He was on duty all night and would ring a bell to signal what animal came to water, one ring for elephants, two for rhinos, etc. The more rings, the rarer the animal. Four rings were for big cats. We could get up to see if we wanted, or we could turn over and go back to sleep. The next morning, we awoke to one of the most beautiful sunrises I have ever seen.

An interesting piece of history occurred at The Ark. England's Princess Elizabeth was on a hunt there in February of 1953 when the word came to her that her father, King George VI, had died and that she was to be crowned queen. One of her mounted heads was on display in the Ark.

The next day on safari, we encountered every African plains animal you could think of. As you know, the animals think vehicles are just another animal and pay no attention to them. It was really interesting to see so many animals up close in their natural habitat. We took lots of pictures and had a lot of fun. Later, dentist Richard Churchman, nurses

Debbie Turner and Mandy Haley, and I walked up a river looking for hippos and came across several Maasai Warriors in full local dress tending flocks. When I tried to take their picture, they would jump behind bushes. We went over to try to talk with them, and with hand signals, they let us know that we must pay to take their pictures. We were glad to do so.

After returning home, Charles and I continued to try to learn more about how the eye works and more about fitting glasses. Charles wanted to know all he could about the eye and eye problems and diseases. I wanted to know those things, too, but I mainly wanted to know more about how to calculate the proper lens for each patient. Every time we returned from a trip, we had more questions for Dr. Rick McGuirt.

San Dino, Nicaragua
The next trip was to the San Dino Zone in Managua, Nicaragua. Jim took a group of ESL teachers there. Charles couldn't go on that trip, so my friend Rick Credeur took his place. I gave Rick a quick course in refracting, and he and I gave out glasses, while the others taught English and held church services.

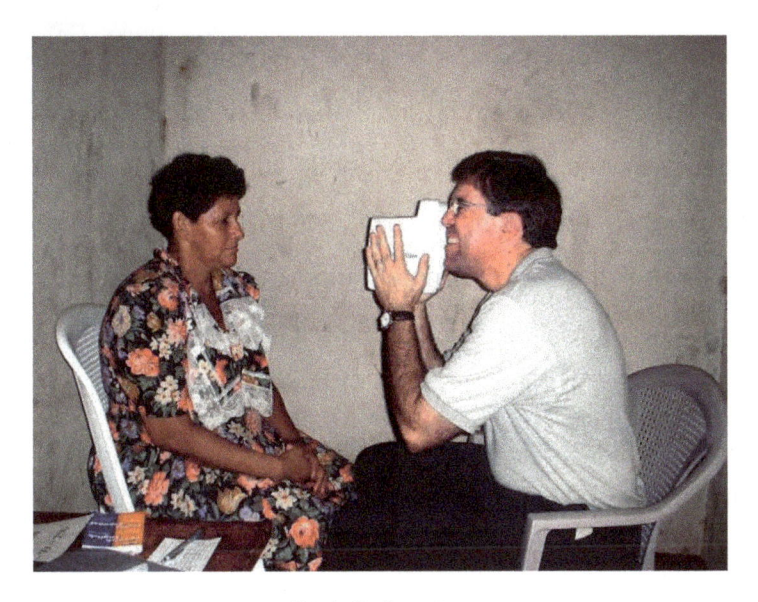

Rick Refracting

The church was in a poor area on the outskirts of Managua. The church building was in a compound surrounded by a high concrete block fence. For security, the gates were kept closed at all times, and there was always someone there to open and close them. Across the side street was a primitive school building with 5 or 6 rooms. We set up our refractor in a small storeroom at the school and our glasses dispensing area on an open porch. That suited me fine because I was out in the open where there was a little breeze, and Rick was cooped up in a hot dirty dark storeroom. The temperature must have been way over a hundred in there.

At San Dino, two sisters from the church, Connie and Alejandra, helped Rick and me with crowd control and wrote names on prescription sheets. BMDMI provided Eduardo and Allan as translators, and Allan also helped make up custom glasses. He was really good. Our cook was Beverly, who had cooked for us on the second El Tuma trip. Pastor Rigoberto and wife Connie (another Connie) have two children, Figueroa, about 20, and his little sister Conniesita, maybe about 10 or 11. You will hear more about them on future trips to Managua. Jim's ESL routine takes two weeks, but Rick and I only stayed for one. Four days of intense glasses work, from can't see to can't see, is enough for us. Rick refracts a little faster than I can give glasses, so he has time for breaks. On the other hand, I worked from early 'till late without so much as a restroom break. We sweat enough in the tropic heat not to have to go.

Jim Selfridge was the team leader for this trip. Members of his team teach English as a second language (ESL). Their class members really do learn a good bit of English, but they also hear the gospel. The first week the teachers make friends with the class members. Toward the end of that week, they give their own testimony, and the next week class members are invited to make a decision for Christ themselves. Besides the class periods, they hold big group sessions and preaching services. Many of the students make a first-time profession of faith. Jim was glad to have us go along and give glasses. We slept in the little church building and then, in the morning, moved all of our stuff away so the church service and group sessions could be held that day and evening. Jim wanted everyone who came to learn English to also hear the gospel. They heard it in the classes and also in fun-type church services. Sleeping was cut short every morning at about 3:45 or 4:00 because about 6 feet from the side of the church was the neighbor's chicken yard dominated by a huge colorful rooster. Jim was especially upset with the rooster. Every morning he threatened to kill it, but somehow it survived our time there. The San Dino trip began in December of 2001 and extended into the next year. At midnight on December 31, we had another rude awakening. I thought I was back in Korea with lots of incoming as well as outgoing artillery and small arms fire. Almost everyone in Managua must have been shooting off fireworks. The sky was filled with rockets, and huge firecrackers were going off everywhere. The tin roof on the church was vibrating so much it was setting up a racket of its own.

I quickly slipped on some clothes and went outside just as little Conniesita stepped out from a little Sunday School building where the family was staying. When she saw me, she let out a whoop and ran and jumped into my arms. We climbed up on the compound wall and watched the show for what seemed like an hour. We learned later that the Nicaraguans prepare meals, sometimes elaborate ones, and set the tables early in the evening. The families and guests get together and visit and dance and sing until midnight when they set off the fireworks. It is only after all that they eat the meal. Those Nicaraguans really know how to usher in the New Year.

On a day off, Armando, Connie, and Conniesita took Rick and me downtown and showed us an ancient Spanish Mission. A Hollywood group was there making a film, and we were not allowed to go in. I wanted to get a picture of the inside of the church, so I watched for my chance and got in long enough for a couple of shots. Some government buildings were shown to us, and we also went to the shore of Lake Managua, where a whole pig was being roasted. Lake Managua is a huge lake and is full of fish, but because of pollution, most people will not eat them.

Another day off activity was a visit to Catarina, a beautiful little village of artisans, musicians, and nurserymen. Beautiful shrubs and flowers were everywhere. They are grown in plastic bags and are available for sale. Villagers can be seen creating works of art, making musical instruments, making music, telling stories, and tending and selling flowers. Catarina is higher than the surrounding areas, and it overlooks a beautiful lake. A cool breeze seems to always blow up from the lake.

The night before we left, Armando took Rick and me to his home because it is close to the airport, and we were to leave very early the next morning. Little Connie (Conniesita) was in a playful mood, and after a while, I struggled through the story of the three bears in Spanish to try to calm her down. Connie speaks very little English, and she listened with rapt attention to every word until I got to the last line when she shouted in English, "…and there she is." When we turned in for the night, I kissed her good night, told her I loved her, and hoped someday to see her again.

Charles and I continued to try to learn as much as we could about the relationship between the eye and glasses. We continued to have a sometimes friendly, sometimes not so friendly rivalry. Charles never passed up an opportunity to try to prove me wrong in my selection of glasses or anything else, but we remained on good terms.

San Jeronimo, Honduras

My old friend Harvey Kieffer, who was my roommate on the Kenya trip, makes regular trips to Central America. Harvey's church, Maplewood Baptist, sends two or three groups to Honduras and Nicaragua each year, and Harvey is the team leader on most of them. After San Dino, Charles and I signed on for a week with Harvey's team in San Jeronimo, Honduras. Besides Charles and me doing Glasses, there were Medical Doctors, Nurses, Dentists, Pharmacists, and people who gave out clothes. We flew into San Pedro Sula, Honduras, where a bus was waiting to take us and our gear to San Jeronimo. It is a small village where most of the ladies are homemakers, and most of the men work on a farm. If you ask the men what kind of work they do, they would say simply "machete." The night before we arrived, a local man who was said to be a drug dealer was shot and killed by police. Wow! Even in a poor village in a third-world country.

The team worked in a school building and held services in a church up the street. I set up a place for us to give out glasses under a large roof overhang, and Charles chased the bats out of a little room without windows and set up his refracting table. Charles had been an instrument man at one of the plants, so he was our "tech" guy. The crowd control people had a little trouble with the people waiting in line. The men refused to wait in line behind a woman, so there was a lot of cutting in. Our crowd control people finally had to get the people to form two lines, one for men and one for women, and they let in a person from each line in turn.

We continued to try to train local people to put custom glasses together, but on this trip, we had only limited success. We did, however, continue to gain confidence in "prescribing" the correct lenses and were gaining speed as well. By this time, we had some cylinder correction lenses and could correct for astigmatism. The idea of showing God's love to the people was really catching on with us, and I especially made sure that I made personal contact with the people and showed them in the short time that I had with them that I cared about them.

We had some extreme cases in Jan Jeronimo. A mother and grandmother brought in a little three-year-old girl who had extremely poor vision. Charles was not able to get any reading with the refractor, so I had to determine the correct lenses for her in another way. She knew the letters of the alphabet but could only read letters that were two inches tall. After a little prayer for help, I found that she could see a little better up very close but not farther away. So she was near-sighted. Through trial and error with negative lenses, the needed correction was found, and she was able to read the very tiny print.

A little older girl had a similar problem. According to her mother, she had always been a happy child, but something had happened to her vision, and she had dropped out of school and become very despondent and withdrawn. It was painful to see the hopelessness on her face. When she was fitted with glasses, a big smile came on her face. Her mother cried for joy and said that it was the first time she had smiled in over two years. That one had me crying too.

Charles had a great translator who filled out the prescription sheets and also did the mathematical calculations. My translator was a dark-skinned man who had a habit of disappearing just when I needed him most, and he was continuously eating candy. I think he was going into the woods for smoke breaks. It didn't hurt too much, though, because I was beginning to know enough Spanish "glasses" words that I seldom needed his help anyway.

The side trip from San Jeronimo was a visit to the Mayan ruins of Copan. The lawns were well kept by crewmen with machetes. Many of the ruins had been restored completely and were amazing in their design. Others were in the process of being restored. They were similar but somewhat different than others I had seen in Mexico.

Dr. Barry Newton diagnosed a little girl of about five or six as having congenital syphilis. She could have been cured with the proper medication, but we did not have what she needed. Barry obtained the girl's address, which was in another village, and when he was back in the States, he tried in vain to get missionaries or local doctors to find and get treatment for her. I tried to help and eventually became almost obsessed with the problem. I contacted BMDMI to get help for the girl, all of our efforts were without success. The BMDMI missionaries were clear across the country from San Jeronimo. I even planned to make a special trip to go back and locate the girl and even thought about bringing her to Lake Charles for treatment. Harvey told me that he had had similar situations and pointed out all the problems involved. There seemed to be no solution. He also said that the people were very reluctant to giving their address and the address we had was probably false. Eventually, I had to give up the idea of finding and helping her, and it broke my heart.

China

Charles had made a trip to China the year before and on a later trip without me he was asked to be a prayer walker. He spent the time there walking around the area where others were doing medical and other types of work. I think that at first, he would rather have been doing something more active, but I could see a real change in him when he returned. He became even more dedicated to missionary work and came to love China. As you know, there are missionaries in China who must operate in strict secrecy, as must Chinese evangelistic Christians. China has strict laws against spreading Christianity. Foreigners who break the laws are deported, but Chinese who do face long prison terms and harsh treatment. In spite of this, there are many underground "home churches" there. Because of this, local missionaries do not give out their phone numbers or email address to Americans or others without their first having learned the rules of communication in China. People sending emails to missionaries in China must never use names of people there or to mention Christ, Lord, Jesus, or words like prayer, faith, etc. In China, it is all right to be a Christian and to worship as a Christian, but it is not all right to tell others about your faith unless they ask. There are even some sanctioned Christian churches for those who already believe in Christ. They are watched closely to make sure they don't evangelize.

Charles came back from his China trip talking about a missionary from Holland that I will call Tru. He also spoke often of a young American missionary that I will call Bre. Time after time, Charles told us about these two ladies until finally, we asked about Bre's family and her background. I finally realized that I knew Bre, and she had once belonged to our church as a young girl and had been in our home and had even invited Steve, my other son, to escort her to her prom. We also knew her parents and had kept up with them from time to time. Neither Bre nor Charles had known of their connection to Trinity Baptist Church. Rissi and I were anxious to see a picture of Bre, but Charles didn't have one. She and Tru did not want their pictures taken. Tru and Bre are good friends and work together when they can. Tru and her husband's work consists of locating people who are hair-lipped or tongue-tied and getting surgery and therapy for them. Bre and her husband work in Thailand, China, Cambodia, and other countries.

In the summer of 2001, and with Tru's help, Charles and I were able to obtain an official letter of invitation to enter and travel in China and to give glasses to people in China who otherwise would not have them. We went to Houston,

TX, to obtain our Chinese visas, packed up our glasses, and took off. We arrived in Beijing after a very long flight and spent the night at the Friendly Hotel. The next morning we flew to Guilin in Western China, where we were picked up by car and taken to meet Tru. After a couple of days' rest, we loaded into Tru's jeep, picked up translators Greg and Helen, and headed southwest into the mountains. Charles is a big guy, probably about six-four and at more than 250 pounds. Because of that, he got the front passenger seat, and Greg, Helen, and I bounced around in the back.

Charles, Helen, Dick

It seems that the higher up in the mountains people live, the poorer they are. Two men and two women followed us in a car. They worked for the Chinese government, and their job was to monitor our activities at all times. I suppose that if we stepped out of line, it would be a big deal. It took us a couple of days to get where we were going first.

China is a fascinating place, and the long drive enabled us to see many strange and wonderful things. The mountains were a sight to behold, unlike any we had seen before. Those I saw along the Li River a few years before were similar but not exactly the same. There were tall, slender peaks, some with spike-like tops and some more rounded. Rice grew on every flattened space, large and small, and all arable land was terraced. It was harvest time. I tried my hand at threshing the rice and had to be given instructions on doing it right. Little boys could be seen herding pigs, water buffalo, and other farm animals through little village streets. One little boy easily herded about 50 ducks along the main street. People lined the streets selling products from their farms. Young girls hauled two pails of water at a time that held at close to 5 gallons on each end of a stick across their shoulders. Little trucks built onto motorcycles carried huge loads. Here in Western China, the people do not wear modern clothing as in Beijing or Shanghai, or any other big city in the east. Many of the people are dressed in the traditional garb of their individual "people group." Ladies also sell beautiful parts of their traditional clothing that they say are hand-made by their old grandmother. Yeah right! They are beautiful, though. I brought some home for Rissi. Buddha and god idols were much in evidence, as altars were just inside people's homes. We saw one little shrine with images of what must have represented parents. They had been sprinkled with blood and parts of chickens, and other animals were left there. In China, there is the worship of Buddha, ancestor worship, and there is also animism (worship of nature like rivers, mountains, etc.). The people were amazingly friendly, especially in the small villages, many wanting us to come into their house and eat with them, even though we could not speak their language.

We worked in schools, giving glasses to students, as well as to professors and some townspeople. Charles told me the first day of work that he wanted me to start giving out glasses, and he would start refracting.

One old lady we saw that day had a very bad eye infection. Dick Landry had given me some drops for this occasion and had warned me to test with one drop in each eye before giving the patient the bottle. I did that and had the lady wait about 20 minutes for any allergic reaction. After seeing that there was none, I reached into my pocket, pulled out a bottle, and gave it to her. That evening I realized that I had given her the wrong bottle. The one I gave her was just plain tears and not what she needed to fight the infection. I was really kicking myself over that blunder, but when we arrived back at the school for a second day's work there, I saw the lady selling her wares in front of the school. I swapped bottles with her and could breathe easier now that she had the medicine she needed.

Early one morning, before we left for our day's work, I took a walk through the village and found some people preparing to butcher some pigs. I had seen little boys walking pigs through city streets. They would lightly tap the pigs with the slender bamboo sticks to steer them in the direction they wanted to go. The pigs were really tame. But here, it was much different. The men tied a rope around the neck of a pig, then took a hitch around its snout and dragged it out, squealing at the top of its voice and dragging its feet. I think that was to get its blood flowing. They would then put it over a rack, slit its throat and catch the blood in a pan. Then they did a strange thing. One man brought out a bicycle tire pump, stuck the tube into the pig's foot, and started to pump. He then went from foot to foot and then to the pig's ear. When the pig looked like the Michelin Man, they brought out the next pig and repeated the process. It was only after I asked a lot of questions back home that I found someone who had seen this done with sheep. He said that the pumped air separated the skin from the carcass and made it easy to skin - smart Chinese.

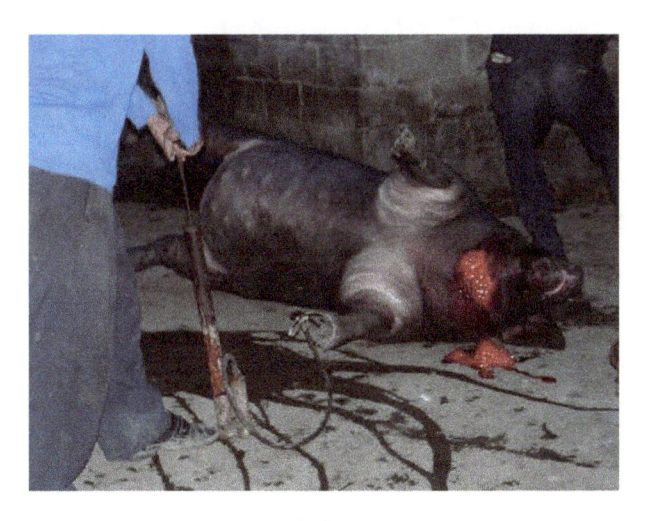

Michelin Pig

We pushed hard, giving glasses and bouncing around on rough roads. One time about mid- afternoon, I could tell that Charles was becoming irritable. He snapped at Tru, making her cry, then began disrupting our dispensing of glasses.

I didn't know what was bugging him but suggested that he go take some pictures and let Tru do the refracting for a while. He did that, and Helen helped me while Tru refracted.

Charles took some pictures and did some prayer walking and was okay when he got back.

We found that the majority of the Chinese we saw were near-sighted, and we were running out of negative lenses for them. At one point, Charles said that he had nothing to give to a school professor. I picked up a plus 1.75 and handed it to him to try. The professor was elated that he could see better, but Charles was deflated. It proved that he had lost the argument that started in Africa.

Helen and I worked together on that trip, and we became good friends. She told me that she had been very close to her grandfather and that all of her grandparents were dead, and she really missed them. I told her that I would be her grandfather, and from then on, she called me grandfather, and I called her granddaughter. I kept in touch with her for a long time by email. She and her boyfriend later moved to Hangzhou near Shanghai and got married.

Typhoon Kate

In Guiling, it was hot and the air conditioner in our hotel room was not working. The girls at the desk could not understand what I was trying to tell them and about that time a lady who spoke English came in and the girls called her over. She called herself Typhoon Kate. Kate, the girls and I all went up to the room. Charles was not dressed when I left so I knocked and called out "Charles, are you decent?" He said "Come on in" and we did. Charles was lying on his bed in his boxer shorts and undershirt. He did not get up.

Now Charles had recently had prostate surgery and was still somewhat incontinent and wore a pad. Lying there in the bed he looked huge. Kate immediately took a liking to him and told him to not worry about being in his shorts and that she was completely Americanized, whatever that meant. Well, we got the air conditioning fixed and for the next three days we met with Kate from time to time for meals and sight-seeing.

As a young girl, Kate had married a very wealthy older man. She spoke English fluently and had traveled all over the world.

Typhoon Kate

When her husband died, he left her well-fixed financially. She said she could have continued to live very well in high-priced Hong Kong, but she chose instead to live in lower-cost Guiling so that she could use some of her money to help others. She had built several schools high up in the mountains where they didn't have public schools and even paid part of the teacher's salaries. Typhoon Kate's next project is to build a bridge, which she has designed, across a river that divides a small village during the rainy season. She named all of the schools she built after her deceased husband. She told me about the gods the Chinese worship and I told her about Jehovah God and about Jesus and how he saves those who believe in Him. I hope she came to believe and I hope to see her again in heaven. I liked Kate a lot.

Charles and I had another treat when we took a three hour bus ride from Guiling to Kaili to visit Lindsay Saucier. Do you remember the girl on the beach in Costa Rica who was trying to decide between a medical career and as a missionary to China? Well, here she is in China.

Lindsay

We had a great visit with Lindsay and her roommate Shelly. Lindsay had learned to speak Chinese very well and boldly told everyone about Jesus and had had several converts by the time we met her there. When confronted by officials, she simply said they should tell their people to stop asking her about her God. Lindsay was overjoyed that a Chinese evangelist named Luke had come to Kaili to join in the work there. Luke was known by the police and had barely escaped being picked up more than once. He was a wanted man and would face harsh imprisonment if he were seen. One day Lindsay, Luke, Charles, and I were in Lindsay's new apartment that she was just moving into when Chinese officials dropped in for a visit. Luke put his head down and pretended to be working on a lock. Then when the officials had their backs turned, Charles grabbed him and hustled him down the stairs and away from the building.

After we had given out all of our glasses, we received commendations and gifts of tea from the officials who had followed us, and we were free to do some sightseeing on our own. We bought gifts at Beijing's famous silk market and shopped for pearls at a fabulous store. Sellers in both places expected to be negotiated down on their prices, and we had fun doing that.

We also flew to Xi'An to see the Terra Cotta Soldiers, which had been discovered by three farmers. There are about

8000 slightly larger than life, clay figures made during the Qin Dynasty (211-206BC), each an exact replica of an actual soldier. If you are not familiar with them, you may want to check them out on the Internet. We had our pictures taken with one of the "famous" farmers. When I was there several years before, I had quickly bought a book with a lot of pictures about the soldiers and had the farmer sign it for me. Then back home I discovered that the book was a Spanish Version. This time I bought an English copy and the same farmer autographed it.

Dick with Farmer

In China, we had no way to tell about Jesus, except to Typhoon Kate. Lindsay spoke the language well and was bold in spreading the Gospel one on one. Tru had been stopped with the work she was doing once and was much more cautious. She said that she was not able to "sow" or to "water" but was "plowing new ground" by showing God's love through her work. Charles later told me that he "shared the Gospel every day in China," but he did it without words.

I could write much more about China and our trip there, but I will end by saying that Charles and I were both very attracted to China and cherished our time and experiences there. Charles has made several trips to China, staying months at a time since then.

Continuation of the Glasses Ministry

That was the last trip that I wrote to Tim about, and I now regret that I did not continue to record them. We continued to make the trips and continued to learn to make the giving of glasses faster, easier, and more accurate. We also learned to make our evangelistic ministry much more effective. The glasses ministry became and continues to be my driving passion.

We later bought a Nikon Retino-max refractor and no longer have to have a darkroom in which to use it. We now only use it on the difficult cases as described in "visionforjesus.org - Learn How in English."

Our glasses ministry developed and grew through the following years, as I learned more and more about how to fit people with glasses. I was helped a lot by Dr. Rick McGurt.

Every time I returned from a trip, I had a lot of questions for him. It was Rick who told me we needed a refractor. Dr. Mel Guerig taught me how to do the calculations to get to the right prescription. When I was a little slow to catch on, he gave me a room in his office and one of his technicians, telling her to "Stay with him 'til he gets it."

Mel also gave me the Spanish reading chart from which we have made thousands of copies that we still use. From that information, I made up the prescription sheet that we use now.

Incidentally, the word prescription is a misnomer. We don't prescribe glasses. A prescription sheet is simply a place to write the numbers from our refractor and to calculate the probable strength needed. We use that information and other things to locate glasses that help the person. When we find one that helps, we offer them stronger and weaker corrections until they find the ones that help them the most. If they still don't see well enough we suspect astigmatism and correct for that.

We had some interesting trips with Pastor Restrepo in Colombia and after the first one I overheard him say about our giving glasses at his church, "It's the best thing we ever did." With him we learned to eat and enjoy capybara. Restrepo was a kind of head pastor over several churches and we great time with one in Villavicenco. After our first trip with him he told me this story:

There was a man who lived next door to the church who hated the church and the people that attended there.

For the twenty years he lived there he harassed the church and its members. He would curse out at anyone who parked with any part of the car in front of his property. He would curse the ones that walked in front of his property. One time he sued the church for singing too loud. He won the suit and the penalty was no singing for a year.

But when we were there giving glasses, someone was brave enough to give the man an invitation to come and get free glasses. He came, got glasses, heard God's word and was saved. Restrepo told me that he had not missed a single service since then and became one of his church's hardest workers. He constantly apologizes for his past actions.

At some point, we worked with Pastor Julio Sanchez in Bogotá, Colombia. Julio became so impressed with the work that he told us he wanted to be on our team. From then on, he and his church members, particularly Javier Carrillo have worked closely with us. They have done all of the logistic work necessary, have found places for us to work in the city, and made sure we had all the local help that we needed.

Dick Using Paddle

Dick Giving Glasses at Paraiso, Bogota, Colombia

For years Julio and his wife Diana put us up in their home. His mother, Bertha, cooked the most wonderful Colombian food for us. We were privileged during that time to live with them and their three girls, Daniela, Alejandra, and Camellia, and to watch them grow from grade-school-aged children to beautiful young adults.

I was making three trips a year, visiting countries in Central and South America. It was getting a little hectic, and I needed some help. Also, by then, I was up into my 80s and wanted to turn the leadership of our program over to someone younger who could continue after my time. The program we had developed was just too effective to let it die with me.

God led me to ask David Self to take the job. After thinking and praying a lot about it, he agreed. He took the job, and he has done a much better job than I ever did. David greatly improved the evangelistic portion of our work and solved an ongoing vexing problem we had in getting in our glasses through Customs in Columbia.

On a typical trip, we now go with a team of three or four from Trinity Baptist Church, Lake Charles, Louisiana, work three long days in each of two Churches and give glasses to some 1,200 people. If we are unable, for whatever reason, to fit a person with glasses, or if we find people with eye diseases or disorders, we send them to professionals at our expense. People over about 40 years of age receive two pairs, one for reading and one for distance. If they ask, we also give glasses for working on a computer. We are able to adjust for astigmatism. Each person also gets a good pair of sunglasses and a Spanish New Testament. Each person receives spiritual counselling one-on-one by one of the local church members and is given the opportunity to receive Christ as their Savior. A big percentage do. David teaches a class to prepare them to do this. He has used "Sharing Jesus Without Fear" and "Three Circles." Our most recent trips have consisted of David Self, Anne Self, Jerry Seal and me.

Masaya

Jim Selfridge was having trouble finding churches in Nicaragua in which to have his ESL ministries. He was working in Managua, and I was finishing up a trip to another city. I told him that if he could furnish me with a car, a driver, and a translator, I would go to Managua and find some churches for him to work with.

So, after I finished my work, I did just that. The car belonged to the pastor of the church in San Dino, the driver was his son, and my translator was Leda.

For Jim's work and for mine, we need medium-sized churches. The really big churches have their own agendas and are not interested in us. Nearly all of the churches were eager to have either of us, and we worked with some of them, but the work was limited because of the shortage of space and the number of helpers the church could furnish.

For three days, we had a great time driving around the city. We would stop for lunch at nice restaurants and get to know each other. Leda was interested in Jim's work and in mine, and we became good friends. We were having a good time but had visited church after church without finding even one that was really right for Jim. Finally, early on the fourth day, after visiting one church, I said, "I quit, this is not working, and I want to go to Masaya." Masaya was a town south of Managua. I don't know why I wanted to go to Masaya. I knew nothing about it, but I was ready to give up on finding a church.

We went to Masaya without even looking for a church, there was a big new round church building with open walls and a sloping floor down to a central platform. If you look at Masaya on Google Earth, the big round church almost jumps out at you. There was a campus there with a school and a big separate cooking and dining area. It was a Baptist Church, and the pastor and church leaders would be glad to have both ESL and glasses ministries at their church.

Isn't that the way God works? He lets us exhaust ourselves trying to accomplish things in our own way, and when we give up, He takes over and makes it happen.

Jim and I each made a number of very successful trips to Masaya, and Greg Bath went there a couple of times to work with young people.

Leda made a number of glass trips with me, working as a translator. She has come to visit me in Lake Charles, and we continue to stay in touch through Facebook.

Leda at Dick's House

Western China Cleaver

Charlie Rue and I were in a little village in Western China where there was an artesian well flowing in the center of the village. All the people came to the well to get their water. I was amazed that young girls would carry two big buckets of water on a pole across their shoulders.

Water Girl

In other places where I had been, old women would carry water on their heads without even holding them. I put one of their buckets on my shoulder, and I would not be able to carry it very far.

Across the street from the well was a little open-air café run by a little lady. She always seemed to have a clever in her hand, and she chopped so fast that I wondered how she could do that without chopping her fingers. I always like knives of various kinds, and I thought the clever would make a very nice souvenir. I talked with Tru, the missionary we were with and told her that I wanted to buy the clever and wanted to pay her many times what it was worth. The lady said that I could buy one just like it down the street for almost nothing. I left the matter with Tru. Later Tru gave the clever to me and told me she would take only a small amount for it.

Then I learned that the clever had belonged to her father, who had run the restaurant until he died. Now I felt really bad and asked Tru what I could do for the lady. She told me that the thing she would most cherish would be a letter from me. She said that the lady would frame it and hang it on her wall. Tru said that she went back to the village every six months and that she would take the letter to her.

I wrote the letter and sent it to Tru. I hope the lady got it. The clever hangs on the wall behind my computer, and when I use it, which is often, I always think of the little lady by the water.

Chinese Soldier

Whenever I was in China, I frequently asked if anyone knew any Chinese men who had fought with North Korea in the Korean War. I had previously been to Korea and China since the war but had never been able to find one. When we were back in Guilin, Tru told me that she knew of an old man who had fought in a war and that it must have been the Korean War. She contacted him for me and set up a time for us to meet.

Tru dropped me off at Wong's home, and we had a long visit. Wong received me into his home and served tea and fruit, including an orange-like fruit the size of an extra-large cantaloupe.

Wong had fallen out of favor with the military after the war and was forced to quit. He then went back to school and to university and became an English professor. He was no longer teaching full time, but because of his knowledge of English, he was called upon from time to time. Wong still lived on the university campus.

We talked about the war and the different ways that Chinese and Americans have of fighting. He said that, by far, the worst thing about the war for him was the bombs. I took that to mean artillery shelling, as well as bombs dropped by air. In Korea, we fired artillery on mountains where the Koreans and Chinese were holed up until you would have thought that nothing could still be living there. Then when we went in, they would come out of their underground tunnels by the thousands.

Wong said that each man carried some rice on his back, and when they cooked, a small amount was taken from each soldier. They never had enough to eat, and the conditions were dreadful while they waited underground. He said that they were more afraid of being wounded than being killed because there was little or no medical treatment for the injured. In most cases, the injured faced painful recovery or slow and painful death.

The veterans of the war held reunions as we do in the U.S., but only one in a hundred of their soldiers were still living; (at that time), whereas the vast majority of the American Vets were still living then. Wong and I had a long and pleasant visit, and we talked about the friends we had lost. There was no animosity between us. Years ago, we would have killed each other. Now we were just two people who had had very different experiences in the same event.

Wong

Dian Dance

We were having more and more problems getting our glasses through customs (Dian) in Bogota. Sometimes they

193

would make us wait all night while they opened every box and every carton. They would dump out each carton count each pair, and carefully put the glasses back in the carton.

On this occasion, I made a mistake on the document I filled out, and they decided to punish me. They would hold the glasses and jack me around for a few days before releasing them to me. I learned of this plan later.

The airport where Dian operated was in North Bogota. We were working with Javier Perilla, who had us set up to give the glasses high up on a mountain in south Bogota, two hours away.

Each day we were sent from one official to another until we were at the office of the guy who would make the final approval. He did not give it. We would not get the glasses.

We had to make one last trip to the Dian office at the airport. Javier was really mad.

When we got there, the Dian office personnel were having a party. There was a small combo band with a pretty singer/dancer and a really good trombonist. There were balloons and streamers and party hats, and everyone was dancing to the Colombian rhythm.

Javier was livid. He said, "They treat us like dirt, and then they have a party? I'm going to give that Dian Director a piece of my mind." I said, "No, Javier, we are not going to do that. We are going to go in there with smiles on our faces, and we are going to apologize to them for causing them so much trouble." "What?" He said, "I don't think I could do that." Then he disappeared. I think he went to have a little prayer meeting with himself.

Finally, he came back, and we went in. The party was going on behind a low railing, and we were on the outside of the rail. I started taking pictures of the whole thing. I took pictures of the combo, the decorations, and the singer-dancer. After a while, one of the girls opened the gate, took my hand, and led me in. I'm not a dancer, so I twirled her around a few times and went back to taking pictures. At one point, I tried to imitate the gyrations that the dancers were making and got a lot of laughs. As I took a picture of one of the girls, she asked, "Was it worth all you had to go through to get your glasses?" I told her that we were not going to get them and that we would bring them back to the States. She went right to the head lady whose birthday we were celebrating, and the birthday girl immediately stopped the music. She talked to some of the others then made a phone call. In a few minutes, the girl who asked me if it was worth it said to me, "You are going to get your glasses, and you are going to get them now, and it's because you danced with us." I looked at Javier. "Did you get that?" Javier was floored. He went back and told everyone, "He's a man of God."

I became Facebook friends with Roger, the trombonist, and with Carmen, the singer; we still exchange messages from time to time.

But that's not the end to the story.

The next time we went back to Bogota, I was really looking forward to seeing some of my Dian friends, but none of them were there. I learned that they had all been fired. The birthday girl had gone over the head of her superior in releasing the glasses to us. Also, I think the party may have ticked off the head man. It broke my heart.

Website

Some 9 years after David took the reins to the Glasses Ministry, our work was still going strong, but our long-term problem had not been solved. We still needed to find a way to keep this program going after we were no longer able to do it ourselves.

Finally, we decided to produce a website, the purpose of which was to inspire others to do this work and to give them all the tools and know-how to do it. We did that, and it's available to everyone. Just go to visionforjesus.org.

Making the website was the easy part. Now the job is to get it to the people who will take it and use it. There are many churches and other groups that regularly send their members out to other countries with various programs to aid people and to spread the Gospel. It is my hope that people involved in these programs and others will read my stories, learn of the Glasses Ministry and start their own program.

Wouldn't it be great to have many groups like ours, going out to various countries of the world, giving glasses and bringing salvation to the lost?

To help bring that about we have established "Vision for Jesus" a non-profit corporation, run by a Board of Directors. Vision for Jesus will give not only instructional information but direct financial contributions to those groups beginning or continuing the Glasses Ministry. Information is available on visionforjesus.org.

Vision for Jesus can only operate on funds received through generous tax free donations from people like you.

Donation Checks should be sent to:

Vision for Jesus

P O Box 7371

Lake Charles LA 70606

United States

Please see our website visionforjesus.org for other ways to donate and for applications for funds to begin or continue a glasses ministry.

Chapter 8: Other

My Friend, Jimmy

Hoop Net Fishing

During a time when I was in business, Jimmy was fishing hoop nets, among other things. He asked me several times to spend a weekend with him on his boat and run the hoop nets with him. I was busy and, for a long time, put him off. Finally, one weekend I did go with him, and I couldn't believe what he was catching! First, we picked up a huge blue channel catfish. Jimmy tied a rope loop through his gills and mouth so we could handle him. Later, when we got to the dock, I picked up the cat by the rope and heaved him over my right shoulder. His mouth was at the height of my right ear, maybe higher, and his tail was dragging the dock. We also caught several big Opelousas cats in the 40 and 50 lb. range and a whole lot of normal-sized cats. I told him, "Jimmy, had I known you caught fish like this, I would have been here with you long before now." But I was busy with my business and never went back. Many years later, Jimmy confessed he had never caught fish like that before or since.

As I was writing this, I wondered how our big blue channel catfish compared to the largest on record and found that the record holder measured 58 inches in length. Wow! Our fish must have been at least 65 inches!

Recently a friend told how she had asked, "Why me?" when given an unfavorable medical diagnosis. Now I am asking, "Why me?" Why have I been allowed to see three fish that were larger than anything ever caught; to help catch one, to wrestle one, and to see another?

The Old Man's Knife

During my high school years, I spent a lot of time at Jimmy's house. It was a small four-room house where his parents raised six children. First, there were three boys and then three girls. Jimmy was the oldest.

Jimmy's dad was a hard worker and an avid hunter, trapper, and fisherman. He always had a garden and sometimes raised hogs and chickens. Every evening after supper, he would sit in the back door of their little house and knit cast nets.

The three boys grew up like their dad, always fishing and hunting. There was Jimmy, the oldest, then Bill and Freddie. Their mom was always busy cooking the game they brought in.

I didn't really get to know the girls much at that time. The oldest was Lana, then Annie, who everyone called Teet, and Oma, who was called Nonie. Later in life, I got to know Lana and Nonie.

I loved being at their house. There was always something going on there. I always enjoyed eating at their table with the family, and they all seemed to enjoy my being there. Their mom was an excellent cook. She would hustle around in the kitchen, put the food out on the table, and while everyone else ate, she would sit in the rocking chair and nurse Nonie, the youngest.

Jimmy was my friend, and we spent a lot of time together, but I also occasionally hunted with Bill and sometimes with their dad.

Jimmy's dad always carried a knife. It could not be called a pocketknife. It was a big heavy folding knife with a deer-horn handle that he kept really sharp. He would pull it out and use it as a tool for many things; to clean fish and game, to cut twine for cast nets, for opening things, etc.

I didn't know it at the time, but this family that seemed so idyllic to me was actually a very dysfunctional family. Later in life, I got to know Lana and her husband, Wayne. After his mom died, Jimmy alienated himself from the rest of the family, except for Bill. Jimmy wanted none of his possessions to go to his family, so in his will, he left everything to me. After Jimmy died, Teet asked me for a book of family names and relationships that Jimmy had. I didn't let her have it because I thought I should honor Jimmy's request. I'm sorry now that I didn't give it to her.

Now I am learning about one sister cutting off ties with another and scratching her phone number from her contact list. I once drove to Boulder, Colorado, to visit Nonie and learned that she had spent a lifetime in counseling and trying to cope. That broke my heart, and I broke down and cried for her.

Before his dad died, Jimmy spent many days just sitting with him at the hospital. One day his dad asked him, "Jimmy, why are you here?" And Jimmy replied simply, "guilt." His dad gave Jimmy his knife.

Jimmy didn't want the knife, so he gave it to me. I kept it for a while, but I didn't feel good about it. It had been such a prominent thing in the family that I thought a family member should have it. One day I brought it to Bill to give it to him.

When I pulled it out, he said, "That's the old man's knife," but he said it in a hushed tone, as if he had just seen the ghost of the old man himself.

Much later, Bill killed himself. He stabbed himself in the stomach twice with the old man's knife. When he didn't die right away, he tried to shoot himself in the heart with an old pistol. The shot missed his heart, and then the old gun hung up, so he tried to strangle himself with a rope. At some point, he died.

Much later, I asked Sheriff Tony Mancuso what would have happened to the knife because I would like to have it again. Tony told me that it would have long since been ground up and disposed of.

Jimmy's Blood Clots

Jimmy has had a couple of bouts with blood clots. At one point after a hospitalization, Jimmy's doctor recommended that he sit in a Jacuzzi. After thinking about that for a while, Jimmy's German blood began to boil. He got really mad. He said, "That doctor knows I don't have a Jacuzzi. I wouldn't sit in a Jacuzzi if I had one". He decided to do the opposite of what the doctor ordered and turned off his water heater. He put a tag on the heater noting the date he turned it off. I didn't know about that until years later.

Jimmy lived in a house that he built for himself on a big secluded lot, and when he no longer had hot water, he quit showering in the house. He began to bathe himself outside using a water hose. He told me later that that cold water hurt so bad he could hardly stand at first, but then he began to like it. He said that after he got used to it, even in the winter, that cold water really felt good and refreshing.

Farewell to Jimmy (Written April 7, 2014)

I lost a long-time friend today and the world lost a truly amazing individual.

Back in the days of youth and strength, we sailed the world together, fished, hunted, camped together and lived off the land. We built our own boats, hung our own nets and caught and sold countless thousands of pounds of fish. Some said he was a little crazy and if he was it only served to make him the special person he was.

Jimmy knew how to do all those things, I did not, but I learned from him. Then we sailed merchant ships all over the world together – What great times those were, sailing the world and visiting foreign ports – heady stuff for young boys looking for adventure.

From Jimmy I learned to build things and became a building contractor. From him I learned the things that helped me not only to survive, but to excel as an Airborne Ranger in the Korean War. From him I learned to fish and hunt and enjoyed a lifetime of that sport. With Jimmy I learned the joy and excitement of travel and traveled much of the world and became a short term foreign missionary. Much of what I am, I learned from Jimmy.

Jimmy spent his last years at Golden Age Nursing Home where he found caring people to take care of his every need. And, yes, he found the love of the beautiful and interesting lady, Ceslie. During those years we ate Georgia Mud Fudge blizzards together and recounted the things we had experienced.

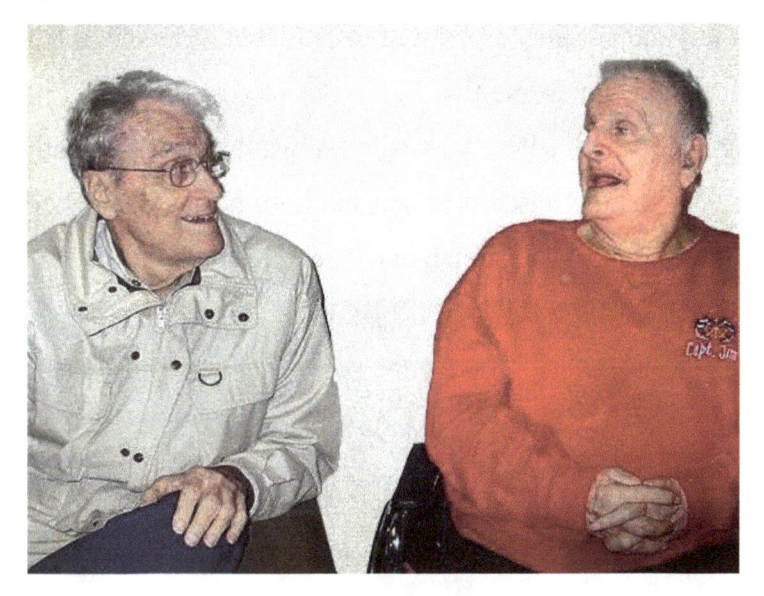

Dick and Jimmy

Rissi and Ceslie

In the last days we talked about faith and what heaven was going to be like. Jimmy was not a religious man and did not attend church, but he believed in God and in Jesus. Jimmy did not pray as most of us pray, but he talked to God and at least one time that I know of God spoke audibly to him. He was hit by an 18-wheeler while driving his little Datsun pickup truck. As he was in the air he heard the words "If you want to live stretch out." He did and survived.

Jimmy and his Datsun pickup truck

At the end he was ready to go. He will be buried in a place he prepared for himself – without obituary or funeral, according to his wishes.

Farewell my old friend, life would not have been so full without you.

Jimmy

Spain with Ted

When Ted Atterbery was living, we were good friends. He had an International Harvester dealership, and when he heard that I was going to rent a jeep to play with for a couple of days, he gave me a brand new red four-wheel-drive International Harvester Scout to use.

Well, I owned some property we called "the Moss Lake Property," or "The Woods," so the first place I went to play with the Scout was on that property.

Actually, it wasn't on the lake because there was a big section of marsh between my property and the lake. There was an abandoned oil-field board road that ran way out into the marsh. An exploratory well had been drilled long before I bought the property, but no oil or gas was found.

I had driven out on the plank road in my car in the past, so after running around on land for a while, I drove out on the plank road.

It had been some time since I was last on the road, and I didn't realize that it had sunk deeper into the mud. When I did notice it, I thought it didn't matter because I was in a four-wheel-drive Scout! Well, you know what happened. When I was nearly at the end of the road, I bogged down.

It was before the days of mobile phones, so I had to walk a few miles to Jimmy Kaough's house to call a wrecker. I told him to bring a long wench line. He did, but it was not long enough. After going back to town for more cable, we finally were able to wench the Scout out.

The underside of the Scout was covered with thick sticky South Louisiana gumbo mud. I used up a lot of quarters at the car wash, trying to get it clean. There may have been some on the underside that was still there when I took it back to Ted. That is not what this story is about, but it shows what a good friend Ted was.

Ted and his wife, Charlene, Rissi, and I all spent a lot of time together. Ted and I shot doves in Old Mexico, hunted deer on his Texas ranch, shot quail in Louisiana, and skied in Colorado.

We read that there were round trip flights to Spain for $300, so we decided to grab that good price and go. When we booked the flights, the special was already over, but we booked them anyway for the four of us and reserved an air-conditioned car. It turned out to be a Porsche.

We had made no plans whatever. When we picked up the car, I got behind the wheel and asked which direction everyone would like to go. We went North along the Castilian coast, then West through the Pyrenees, then down through Madrid, to Valencia, and back North to Barcelona.

The Spanish government had bought up a lot of really large beautiful homes and other buildings all on spectacular sites and turned them into Paradors. They were great. Newly refurbished buildings on sites with wonderful views turned into hotels - and really inexpensive. A room for two went for about $15.00 a night. What a fantastic deal!

We stayed a few days at Benicarlo' at an inn on a little bay with a pebble beach. Our room had a wide doorway (we never closed the doors) looking out over the beach. At night we went to sleep to the sounds of children laughing and playing on the beach. We awoke in the mornings to the soft sound of waves washing up onto the pebbles. What an enchanting place!

Early the next morning, Ted and I went for a walk on the beach. When we passed a couple of girls walking toward us, Ted asked me if they were wearing bikinis. I said, "Not unless they are made of fur." It turned out it was a clothing-optional beach. Men and women and boys and girls were running or playing or sprawled out on the beach with nothing on. Some of the women wore bikinis, and some of the men wore Speedos. After a while, we noticed that some of the girls giggled when they saw us. Our big, baggy shorts looked horrible there, so we both got Speedos. Ted's kind of got lost between rolls of fat, but we came to be very comfortable in them.

One evening a very dressed-up lady came into the restaurant looking down at everyone. She had this "better-than-thou" look on her face and acted accordingly. I thought, "Lady, you've got nothing to look snooty about; I've seen you on the beach."

After we were back home, Ted and I went for a ride in his boat. We didn't plan to see anyone, so after we got out into the channel, we stripped to our Speedos. You know what happened. Our "friend" Ralph Hays came by in his boat and not only got a good laugh but shared the event with others. It wasn't long before it seemed everyone we knew had heard about our boat ride in Speedos.

We all went to a corrida de toros (bullfight). I had been to several in Mexico, so I knew a little about them. Did you know that tickets for the corrida in Spain are sold only to Spaniards? Ernest Hemmingway, who lived in Spain, was fluent in the language and probably looked Spanish decided he was going to buy one. He tried all day, but he failed. Foreigners have to buy them from scalpers.

I asked and was allowed to see the bulls before the fight began. Before the fight, the bulls are very calm and sedate. There can be no noise when visiting. We walked on planks just over the heads of the bulls. They probably didn't even know we were there.

Before the bulls are sent into the ring, they are poked and prodded to get them as angry as possible. They have never before been treated with anything but kindness and gentleness. When the door to the ring opens, they roar into the arena, looking for blood.

A bullfight consists of six fights, with six bulls fought by three matadors. The first fight this day was very unorthodox. The matador (bull killer) did the spit on the ground with his back toward the door. The matador used his cape to deftly lead the bull around him. So far, so good, but the bull did not run on past the matador. He immediately spun around and was towering over the matador just inches away from him. I knew he was a goner, but he took control of the bull, and the whole run was masterfully done. There is a director who directs the whole event and who rules on the bravery of the bull. Charlene was not at all sure about this bullfighting business. She cringed and looked away at first sight of blood, but in a little while, she was jumping to her feet and shouting, "Ole!" with the best of them.

A while back, I read that the bullfight opponents had won out and that there would soon be no more bullfights in the world. Mel and I started making plans. I wanted her to see one, and she wanted to see it. Before we could book our flights, we found that we were too late. It had already happened.

It was a sad thing for me. The bullfight is a beautiful thing with all its pomp and ceremony and strict rules for every part of the event. Every Sunday afternoon, the matador dresses in the traditional Traje de Luces (suit of lights) and performs for his adoring fans and spectators. The bull gets to show his bravery instead of being killed unceremoniously in a packing plant. When the people yell, "Ole!," which means "bravo," it is the bull they are yelling for.

The fight is controlled by the director, who signals the orchestra, which changes the music to let everyone know when one part is to end, and the next is to begin. The director also rules on how the bull is rated for bravery. When the bull is particularly brave, the people wave white handkerchiefs to ask for an ear. If the director allows it, an ear is cut off and given to the matador. It is not just an award for the matador; it is an award for the bull. One-eared bulls are usually mounted with a head mount. I was once fortunate enough to see a fight where the bull was awarded an ear. When the matador is awarded two ears, the head is definitely mounted and displayed. I have seen several of them, the mounted heads. The fight and the bull are remembered and talked about for decades like exceptional football plays are remembered here.

The bulls are bred for their strength and bravery and are treated kindly all of their lives. Only at the end are they provoked to fight and show their bravery. How much better than the ones bred for their meat and unceremoniously killed by a blow to their heads.

It was sad for me to see an end to a tradition that traces it's roots back many centuries. Some say almost back to the time of Christ.

But just recently I learned that there are a few, a very few, places in the world where the traditional bullfights are still held. Mel and I are making plans.

Miss Rissi

When my family was young, I bought a wooden hull boat with twin 25hp motors on it from Mr. Orvis. He owned a tractor dealership in Welsh, LA, and was an avid goggle-eye fisherman. His daughter was my Aunt Harriet. Mr. Orvis had built the boat from a kit and named the boat "The Gulfer."

That boat became a big part of our family's life. We trailered her to various bodies of water around Southwest Louisiana and explored many rivers, lakes, and bayous. We used it for picnics on the beach, for swimming parties, and for fishing. The kids all learned to ski behind the Gulfer. It was an open boat except for the covered bow. The kids liked to crawl up into the covered bow when it was cold or raining.

I kept the boat in my backyard covered with a tarp. One spring, when I started to uncover the boat, I found that the tarp had fallen down inside the boat closing up the bow. I also learned, very painfully, that it was full of wasp nests. Wasps were everywhere in there. It's easy to kill wasps on a wasp nest with a can of insect spray. You just start the spray as you approach the nest, and as you get closer, the wasps fly away, and you remove the nest. That wouldn't work in this case because while killing one nest; I would be vulnerable to the others. What to do? I took a full can of Raid wasp spray, inserted it between the tarp and the side of the boat, and sprayed, trying to fog the entire bow. Every wasp under that bow lit on my hand, crawled up my arm, and flew away.

After several years of use, my kids started telling each other where not to walk on the deck because it was getting rotten. Then one day, I stepped on the gunwale and almost broke through. It was time to retire the Gulfer, and we did.

We all went to a boat show in Houston, Texas, and selected a fiberglass hull which was made by Newman in Miami, Oklahoma. It had a 100 horsepower motor on it. We christened her the Miss Rissi. Everyone knows my wife Laurissa by her nickname Rissi. She was named after her grandmother Laurissa Moffat who had nine boys and three girls. Her boys all called their mother, Miss Rissi. The name seemed fitting.

We used the Miss Rissi as we had the Gulfer, but now we had a boat that we could use to fish in the Gulf of Mexico. I had an extra gas tank built under the floorboard that greatly extended her range. We would go some 50 miles out into the Gulf to fish around the oil well platforms. We would take her through the Intracoastal Canal to Jack Tar's restaurant in Orange, Texas, where there was a dock for diners arriving by boat. One time we took her to Galveston, Texas, where we stayed for several days and fished the Gulf. The kids skied all the way there and back.

Steve, Tim, Mel, Dick

Steve, Mel and Tim

Once I took the kids on a fishing tournament in the Gulf, and Steve caught a prize-winning bonita and got his picture in the American Press.

Twice a year, on the opening day of both the spring and fall shrimping seasons, we would go shrimping in Big Lake. I had built a removable culling board across the stern of the Miss Rissi and bought a 40-foot shrimp troll. We would put the net out and troll for 30 minutes to an hour. When we pulled it in, we dumped the contents onto the culling board. Rissi and the kids would push trash fish and crabs off the board and into the water and put the shrimp on ice. We would shrimp until we caught 200 pounds, the limit of our freezer space. Some days the shrimping was good, and we got our 200 pounds quickly. Other times it was not so good, and it took almost all day.

Sometimes the kids would want to drive the boat while we were trolling, but I told them that it takes a very special skill to know where to troll. I did that because I knew that if they drove the boat, I would have to do the culling. After a few years, they caught on to what I was doing, and I had to let them drive.

We continued to use the Miss Rissi for many years, but then one by one, the kids began leaving home. Steve was married and living in Alabama. Melanie was an RN and nursing in Houston, Texas. Tim was managing a restaurant in Baton Rouge, LA. I was busy with other things, and the boat was not being used.

I went to the Salvation Army and offered them a deal. I would donate Miss Rissi to them and then buy it back for $200. They took the deal. I gave them the $200 and reported the value of the donation as $4,900 on my tax return.

I didn't know what the boat was worth, but I thought that it was worth at least that. I also knew that a donation of a single item valued at 5,000 or more would have to be appraised, and I wanted to keep it simple. Then I gave the boat to my son Steve who took it to Alabama, where he used it for years.

We all love to talk about the great times we had on those two boats. Melanie and I sometimes think we should buy another boat like the Miss Rissi.

Independence Pass

Rissi and I were driving through Colorado when we stopped at Independence Pass. We met another couple there, and we took pictures together.

There was a lot of ice and snow at the high altitude. As I drove down the west side, the road became clear, and I picked up speed. It was clear, except for an occasional icy patch. Rissi asked me to slow down, but I didn't.

There was one of those patches in a right turn. I was having a good time looking at the scenery and was just a little too late starting the turn. We lost traction, spun around, and started heading for the cliff backward! This was before seat belts were in cars. I said, "I'm sorry, Rissi," as I looked back over my shoulder to see how far down we were going to fall. It was a long way down. This was it; we were headed toward certain death.

The back wheels left the road, then the front wheels left the road, and the car was almost vertical when, all of a sudden, the back bumper hit a rock right in the exact middle of the back bumper. We stopped and hung there at an unbelievable angle.

I knew the car could slip off the rock at any time, so we very carefully got out - she on her side and me on mine. It was a long step down to the ground, and Rissi sprained her ankle getting out. We climbed up onto the road just as our friends from the pass stopped to help. They took Rissi to Aspen to call for a wrecker (this was before mobile phones), and I stayed with the car.

The wrecker was able to get the car back onto the road, and I drove it to Aspen, picked up Rissi, and we were on our way again.

There was a small dent in the back bumper made by the rock that stopped us. It was right in the middle of the bumper and only on the bottom half of the bumper. What a miracle that it stopped us at all. I knew without a doubt that God was in control of my car and saved us that day.

Lady at the Drugstore

As I walked into the drugstore, I noticed a lady standing there outside. I thought it was odd. People are either going into a drugstore, walking to their car, or smoking. She was doing none of those things.

After making my purchase, I was in my car about to start it when I heard a tap on my window. It was the lady who told me her boyfriend had left her and asked if I would give her a ride home. Of course, I'd give her a ride.

When I asked her which way, she pointed me south on Elliot Road. She had me turn left on Gauthier Road then north on Ryan. "Why did you have me drive south and now north?" "I thought my boyfriend might be at his boss's house." We continued north on Ryan. "Lady, I think you are giving me the runaround." "I'm not giving you the runaround." By then, we were downtown. "How far are we going on Ryan Street?" "To the end." "We are not going to the end; the end is at the river." Now I'm on alert. She had me turn right on Belden Street. Soon she started looking around - mostly behind us like she was trying to find her way. Another turn or two, and we were on Cherry Street. Now I am on high alert!

Cherry Street is where you go to get illegal drugs. Cherry Street is where you go to find a flophouse where you can get stoned out and sleep it off. By this time, it is raining. Finally, she said, "Just put me out." "No, I'm not going to put you out in the rain." I started heading back toward town.

She had been trying to reach her boyfriend on her mobile phone the whole time, and he finally answered. She said this old man had given her a ride. I thought, "Who are you calling an old man?" I said, "Give me that phone." He said he was near and told me where we could meet. "I'm not meeting you anywhere, buddy. I will drop her off at a safe place and then tell you where she is." I let her out under the canopy at

Walmart Neighborhood Market, where she gave me a hug and a kiss and got out.

It was an intended mugging gone wrong.

Are you my Daddy?

Rissi got a call one day from a man who was looking for his father, Richard Dean Hyatt, whom he had never known. He said that he did not want to call me at my office and embarrass me there. So instead he calls my wife at home? What kind of crazy thinking was that?

Anyway, Rissi told him that it would be okay for him to call me at my office and that if I were his father he would be a very lucky boy.

How about that for an answer to something that could have potentially been life changing. I was very proud of her.

Anyway, I knew there was no way I could have been his father and it turned out he was looking for a younger guy who was an airplane mechanic.

But what an answer. Way to go Rissi.

Train Trip Across Canada

Rissi and I planned a trip to Canada. It was a trip across the country from the Atlantic to the Pacific on a train. The rail cars had glass on the sides and top for complete viewing.

They make many stops where people get off and see the sights.

So, the day came when Rissi and I loaded up in my car, and we headed to Houston, Texas, where we would catch a flight to Canada.

On the drive to Houston, this thought came to me: "I have no business making this trip." Wow, where did that thought come from?! We had been looking forward to this trip for months. Anyway, we went on. After we had checked our luggage and were waiting to board, Rissi told me that she had a bad feeling about the trip. I told her that I did, too.

We went off by ourselves and had a little prayer meeting. After that, I asked, "How do you feel now?" She said, "The same." I said, "Me, too, we're not going." She wanted to know about the money we had paid for the trip, and I told her the money had been spent; the only question was where we were going to spend the next couple of weeks.

I went to the lady at the desk and told her we had decided not to go and asked if we could get our baggage off. She was flabbergasted. She wanted to know why but didn't know how to ask. She finally said, "If you are not going, we have to get your checked luggage off."

So, we went back home. No, we did not read about the plane crash, and we were never sorry we didn't go. The moral of the story is: Pay attention to your feelings. God may be telling you something.

Dick's Right Kidney

I had a going construction business, and things were working well. I had set up a program where my key personnel and I had medical and life insurance benefits. The more the employee was paid, the more life insurance he would get. Of course, I had the most because I made the most. It was a way to charge off my medical and life insurance and reward my top employees. Well, at the end of a successful year, everyone got a raise. The raise kicked me into a higher bracket and qualified me for more life insurance. When my agent asked the insurance company to raise my coverage, they made a routine check on my health. They sent me a little bottle. I was to pee in it and mail it back to them. I did that, and they sent me another one. I sent that one to them and was denied the coverage. I said, "Okay, I didn't need the insurance anyway."

Then I decided that I didn't like that and asked my agent to get it from another company. He applied, and I was sent another little bottle. Then the company sent me another one. I'm thinking, "Enough with the little bottles!" Dana filled out all my forms, so I asked her to start filling the bottles for me. She wouldn't do it.

Well, when that bottle was sent in, the company guy told my agent to tell me that I should go see my doctor. All the usual tests were run, and the doctor told me I had a mass in the lower part of my right kidney. I told him that I wanted a second opinion, and if it was the same, I wanted to go to Houston. He said, "Good! You will get your second opinion in Houston." Well, I told him I had a fishing trip planned to Lake Guerrero in Mexico in a month and that I would go after that. He said unequivocally, "NO, you are going NOW!" and he handed me the pathology report and the x-ray.

I read the report, which said the mass was in the lower part of my right lower kidney. I studied the x-ray really hard but could see no difference in the upper and lower kidney. Something funny is going on here.

The doctor in Houston came into my room while I was still dressed. He said that he was going to cut me from here to here, pointing to the left and right sides of my abdomen, open me wide, and carefully lift the kidney out so as not to spill a drop. I told him, "You are not going to do anything until you look at this x-ray and show me the mass. He lifted it up toward the window for a split second and said, "Oh gosh, yes - It has got to come out!"

I said, "You have got to show me, Doc. I read in the pathology report that the cancer was in the lower kidney. I cannot see anything in the lower kidney." He said, "Do you see the thin little white strip at the top of the kidney? That's the only good kidney you have left!"

So, I had the surgery, and the second day after, I was getting tired of the bed and wanted out. I was able to get a pass, so Rissi and I went shopping. All went well for a while, and I was getting a little tired and was ready to go back. Rissi wasn't through shopping, though, so I waited. I was getting weak, so we went to look for something to eat. It was getting late, and not much was open. We went to the Shamrock Hotel. The waitress took one look at me and said, "What you need is a hot-buttered rum," and she brought me one. Boy! That really did the trick. I was warm again, and I felt like I could go for the rest of the night.

Well, I made my fishing trip and was proud of the picture of me with my shirt off and with my long red scar shining in the sun - and me holding a really big bass!

Motorcycle Ride Home

I helped Tim move from Louisiana to Greensboro, North Carolina. We loaded all of his stuff into a U-Haul truck, and then I drove my motorcycle up a plank into the truck. After we had Tim settled in, I rode the motorcycle home.

I took the Blue Ridge Parkway. It was November, and the Fall colors were spectacular. Every time I rounded a bend, the scene seemed more beautiful than the last. I stopped so many times to take pictures that I was not clocking many miles. After three days, I called home, and Rissi asked if I was in Louisiana yet. I laughed and told her I was still in North Carolina.

Timing the Market

Years ago, I tried my hand at timing the market with some of my mutual fund investment money. The experts all advise against that, but I did it. Sometimes I sold at a nice profit, and sometimes I sold at a loss, but my profits were always better than my losses, and I did well. I only did that for a few years. Much later, after I was retired, I decided to try it again. This was my rationale:

We are still in the age of technology, right? Aggressive investments in technology have been and continue to be outstanding over the passage of time, right? Periodic downturns keep long-term returns from being even better, right? If one were able to be out of the market during those downturns, returns would be fantastic, right? How hard could it be to just sell during the downturns and buy back at a lower cost? Well, you know it is harder than one might think, but I tried to do it and was doing well.

One day when I was trying to determine what to do, I asked God for direction. I told myself while I was doing it that God doesn't tell people how to play the stock market. But you know what? This time He did. I "heard" His voice. It was not His "still small voice." It was loud and distinct. He said, "PUT IT ALL IN AND LEAVE IT." I didn't know if He just wanted me to spend my time doing other things or if He wanted to test my obedience through a recession, but regardless, I put it all in, and it will be there until He tells me otherwise. So far, it's been outstanding.

There is no doubt in my mind now that God "speaks" to me but not always in the same way. I don't think it's just me. I think maybe He "speaks" to you too.

Alaska Bear Hunt

My cousin, Jane Hays' husband, Southey Hays III, is an avid hunter and loves to hunt in Alaska. Southey invited me along on a trip.

We flew to Ketchikan, where we met Don Ross a pilot friend of Southey's and told him we wanted him to fly us to a place where we could hunt black bears.

On a big wall map of that region, we pointed out a big blank space where nothing was shown and asked what was there.

He responded with one word - bear.

We asked him to put us down there and come back in three days to move us to another location where we would hunt for three more days.

As it turns out, we hunted in or near tidal flats, and even though we were inland, we had to carry tide charts of the area. We had to count the number of shallow creeks that we crossed and return across them before a high tide turned them into deep, swift-running rivers.

We told Don to leave word with others where he left us so that if anything happened to him, someone would know where we were.

Don flew us in his seaplane to a likely spot and dropped us off.

There was lots of bear dung on the only high ground we could find, and we made camp there and hoisted all of our eatables high up in a tree.

We had an interesting and fun trip, and I learned a lot about that part of Alaska. It was spring, and the bear were hungry after hibernating throughout the winter. We hunted separately, bringing whistles with us so we could find each other. We each shot a black bear. Southey's male bear had just mated.

I spotted a bear across a wide grassy flat near some trees, but it was too far for a good shot. There was a little creek running in the bear's direction, so I got down and followed the creek to a spot where I could take him. I drew a bead, but before I could fire, he went into the trees. I waited for a long time, and it was starting to get dark. It was now or never. I got up and walked to where I had last seen him when a couple of ring-neck geese came in to roost and lit honking between the bear and me. That spooked him, and I saw him wheel and run further into the trees.

I was able to get off one quick shot.

Now it is really getting darker, and I might have a wounded bear in the woods. Wow! I couldn't leave him, so here I am, going into the fast darkening woods looking for a wounded black bear. Sure enough, I had hit him, and he charged me from about fifty feet away. We have all seen in the movies where a big game hunter shoots a charging animal and drops it just in front of him with a single shot. This was nothing like that. I put three quick rounds into him and dropped him about twenty feet in front of me.

It was 10:00 pm when I shot him and midnight when Southey and I got him skinned out. Two days later, bald eagles had surrounded the carcass and picked the bones clean, leaving a shallow ditch around it where their claws had dug in.

We cut back-straps out of the bear and cooked it at our campsite.

Not bad. We also brought a back-strap home and had a meal there. It was better when we were hungry at camp.

Southey made a full standing mount of his bear and, being a dentist, put a gold tooth in it. It stood in the lobby of his dental office for many years. I made a head mount of mine, and it still hangs on my glassed-in back porch.

Dick's Bear

The Secret of Longevity

A while back, I was in court dealing with a title problem on a piece of land. My attorney, Jim Watson, pointed out our position to the judge. The opposing attorney pointed out the opposing view. I was called to the stand, was sworn in, and was asked a number of questions. One of the attorneys asked the judge if he had further questions for me. He did not.

In further proceedings, Jim mentioned my age. I think I was 91 at the time. Later, when I was no longer on the stand, the judge said, "I do have a question for Mr. Hyatt. What is the secret of your longevity?"

I got to my feet, went back up on the stand, and said, "If you'd like to know, I'd be glad to tell you", and I told him the story. "Years ago, I took stock of the things I had to do in my lifetime: I had to outlive my wife because I could get along better without her than she could without me. I had to keep my businesses going until late in life, especially the one in which my two brothers had an interest because they needed the income, and I needed it as well. I had to sell my businesses late in life to provide cash for my brothers and my heirs. I had to find someone to take over my glasses work because the program is too good to let it die with me, and there were some other things. So, I asked God to keep me healthy while I tried to accomplish all that, and I have been healthy all my life."

So, I guess my story is recorded in the court documents, but Jim wasn't buying it. He pointed out to the judge that I was not the same kind of Christian that both he and the judge had been in earlier days; that I had not indulged in drinking as they had.

Dick and the Girls

When I go to Central and South America, I often work with girls who tell me they want to marry me or live with me. When Rissi was alive, I would tell them that I didn't think my wife would like that. These were beautiful young girls. After Rissi died, the offers came more often. Some would come out and say what they wanted. Others would just become very friendly.

It was usually on the last day after our glasses work was done that the girls would reveal their desires. More than once, there would be two who would become very friendly. Someone in the crowd would call out, "Do you want to marry one of them?" The locals could all see it. That was often the first I would know what was going on.

One young girl's father told her I might be a little old for her. She said "Yes but he's sooo sweet. Hear that guys? If what you are doing is not working, try being sweet.

Once a girl seemed interested and introduced me to her parents who seemed to like me. Someone in the crowd called out as before. She was beautiful and soft and warm. She was smart and spoke English fluently. I thought it would have been like heaven for me. But she was so young and so small. I had to pass. I did it poorly and upset her. I hated that.

This happens more in other countries than it does here, but it happens here as well. Just a few weeks ago, another young girl made her wishes known. Where were these girls when I was young and available?

I once met a lady who was a little older, and I thought, "I could really go for her". We worked together but not close, and for several days we had brief times together and made goo-goo eyes at each other. She made it obvious that she liked me and I wanted to tell her that I really liked her and that if it were not for the many miles between us, I would like to get to know her better and maybe even pursue a relationship with her. I used a translator, which wasn't even needed and did more harm than good, to make sure I was not misunderstood and tried to tell her all that. The translator messed it up; she didn't get the big "if", and it all turned out wrong.

I recently met a beautiful, maybe almost old enough girl in Houston, Texas and was immediately attracted to her. It was not just her outward beauty; there was something more. When we talked later, the attraction grew. She is a Christian and a lovely person. She showed me by running to greet me, by hugs and hand holdings that she liked me too, but we are both too busy for a long distance relationship. Our paths will cross once every two months. Maybe we can be '10 minutes every two months' buddies. We'll see.

Your Body Does What Your Mind Tells It To

Capt. Spragins taught us that, and I have never forgotten it. It has stood me in good stead all of my life. For one thing, it helped us to endure the rigors of Ranger training and of combat.

We all often hear stories about people who exhibited unbelievable strength when it was needed, like lifting a car off of someone who was pinned under it - just examples of the mind telling the body what it must do.

Many years ago, I had a pain in my rear end. It finally went away on its own, but later it came back. So I went to my doctor, who told me I had hemorrhoids. He wanted to cut them, but I wouldn't let him. I said to myself, "I WILL NOT have hemorrhoids. I may have other disorders, but this is my body, and I WILL NOT have hemorrhoids." I have never had hemorrhoids again. It turns out your body will not do what your mind tells it not to do.

I was outside in the sun a lot in my younger life. Consequently, I see my dermatologist, Dr. Michael Cormier, often to have many pre-cancerous spots burned off. Michael tells me that many of his patients think the process is extremely painful. I tell Michael, "You can't hurt me with that little swab," and in the process, I am telling my body, "This will not hurt you." Guess what? It doesn't hurt.

I never let my dentist give me novocaine shots. When the work is done, I want it done; no deadened mouth, bit lips, and inability to talk, right? I once had three root canals and caps done at the same time without any novocaine.

As I write this at 94, a lot of my elderly friends have pain in their hands and knees and elsewhere. When I feel a little pain like that, I refuse it and refuse to slack off on the activities that caused the pain. Soon any pain is gone.

I believe fear of pain serves to increase the pain, whereas boldly refusing the pain makes it go away. Rissi suffered from pain most of her life, and Melanie has terrific headaches. If I regret anything in my life, it is that I was never able to persuade them to use their minds to stop the pain.

Vision for Jesus

Please visit our site at visionforJesus.org to find several things:

1. A website designed to inspire others to begin and continue a glasses ministry like ours and to give them all the information about how to do it.

2. Applications for contributory funding to begin and continue your ministry.

3. Ways to donate to this ongoing work.

Glasses Ministry well done is not inexpensive. On a six work day trip four glasses givers typically give some 3,000 pair of glasses to about 1,000 people. We give glasses for reading, glasses for distance and if requested glasses for computer use. Many of these glasses are the more expensive "special glasses" that correct for astigmatism.

We sometimes are able to also provide New Testaments in the language of the people we are serving. When we find people are unable to fit with glasses or when we find eye diseases or disorders we send them to eye-care professionals at our expense. All of this requires generous donations from people like you.

Vision for Jesus is run by a dedicated self-perpetuating board of directors who I pray will keep it going until the Lord comes again.

Afterword

I hope you have enjoyed reading about my experiences and passions as much as I have in writing them. It has been a journey for me, and I am closer to God now than when I began. I hope you are too.

I first thought I would share these stories with my family and a few close friends. Now because of the Christian message, I want to share them with everyone. I want to share them with the world and am having them translated in many languages.

If you liked the book, please share it with your friends. An e-version of the book is available on Kindle and the printed book on Amazon.

Skip Dent, teaching from the book of Philippians, told us that Christians are a light in the world and that we should put our light out there for all to see. Okay, Skip, here is my light, such as it is, openly shared for all to see.

Tim, Steve, Rissi, Dick, Mel

CPSIA information can be obtained
at www.ICGtesting.com
Printed in the USA
BVHW061257251022
650239BV00003B/59